PATHWAYS
to College Mathematics
GUIDED NOTEBOOK

HAWKES LEARNING

Editors:
S. Rebecca Johnson,
Barbara Miller

Project Manager:
William McCullough

Lead Developers:
Doug Chappell

Developer:
Vincent Cellini

Creative Director:
Tee Jay Zajac

Designers:
Robert Alexander,
Trudy Gove,
Kenneth Hanson,
Patrick Thompson,
Tee Jay Zajac

Cover Design:
Patrick Thompson

Composition
QSI (Pvt.) Ltd.

VP Research & Development: Marcel Prevuznak

Director of Content: Kara Roché

HAWKES LEARNING

A division of Quant Systems, Inc.

546 Long Point Road
Mount Pleasant, SC 29464

Copyright © 2020 by Hawkes Learning / Quant Systems, Inc. All rights reserved.

No part of this publication may be reproduced, stored in a retrieval system, or transmitted in any form or by any means, electronic, mechanical, photocopying, recording, or otherwise, without the prior written consent of the publisher.

Printed in the United States of America

10 9 8 7 6 5 4 3 2

ISBN: 978-1-64277-120-6

Table of Contents

Preface iii
Math Knowledge Required for Math@Work Career Explorations vii
How to Use the Guided Notebook ix
Strategies for Academic Success xiii

CHAPTER R

Review of Foundational Math Skills

R.1 Exponents, Prime Numbers, and LCM 3
R.2 Fractions (Multiplication and Division) 9
R.3 Fractions (Addition and Subtraction) 14
R.4 Decimal Numbers. 19
R.5 Bar Graphs, Pictographs, Circle Graphs, and Line Graphs 25

CHAPTER 1

Algebraic Pathways: Real Numbers and Algebraic Expressions

1.1 The Real Number Line and Absolute Value 33
1.2 Operations with Real Numbers 38
1.3 Problem Solving with Real Numbers 43
1.4 Square Roots and Order of Operations with Real Numbers 47
1.5 Properties of Real Numbers 51
1.6 Simplifying and Evaluating Algebraic Expressions. 56
1.7 Translating English Phrases and Algebraic Expressions 59

CHAPTER 2

Algebraic Pathways: Linear Equations and Inequalities

2.1 Solving One-Step Linear Equations 65
2.2 Solving Multi-Step Linear Equations 70
2.3 Working with Formulas. 75
2.4 Applications of Linear Equations 78
2.5 Ratios, Rates, and Proportions. 81
2.6 Modeling Using Variation. 86
2.7 Solving Linear Inequalities in One Variable. 90

CHAPTER 3
Algebraic Pathways: Graphing Linear Equations and Inequalities
- 3.1 The Cartesian Coordinate System, Scatter Plots, and Linear Equations 99
- 3.2 Slope-Intercept Form 108
- 3.3 Point-Slope Form 115
- 3.4 Introduction to Functions and Function Notation 121
- 3.5 Linear Correlation and Regression 127
- 3.6 Systems of Linear Equations in Two Variables 132
- 3.7 Graphing Linear Inequalities in Two Variables 138

CHAPTER 4
Algebraic Pathways: Exponents and Polynomials
- 4.1 Exponents 151
- 4.2 Scientific Notation 157
- 4.3 Modeling with Exponential Functions 159
- 4.4 Addition and Subtraction with Polynomials 164
- 4.5 Multiplication with Polynomials 169

CHAPTER 5
Algebraic Pathways: Factoring and Solving Quadratic Equations
- 5.1 GCF and an Introduction to Factoring Polynomials 175
- 5.2 Factoring Trinomials 179
- 5.3 Special Factoring Techniques and General Guidelines for Factoring 184
- 5.4 Solving Quadratic Equations by Factoring 188
- 5.5 Operations with Radicals 192
- 5.6 Solving Quadratic Equations by the Square Root Property and the Quadratic Formula . . 198
- 5.7 Applications of Quadratic Equations 202
- 5.8 Graphing Quadratic Functions 206

CHAPTER 6

Geometric Pathways: Measurement & Geometry

- **6.1** US Measurements 217
- **6.2** The Metric System: Length and Area 220
- **6.3** The Metric System: Capacity and Weight 223
- **6.4** US and Metric Equivalents 226
- **6.5** Angles 230
- **6.6** Triangles 239
- **6.7** Perimeter and Area 249
- **6.8** Volume and Surface Area 260
- **6.9** Right Triangle Trigonometry 267

CHAPTER 7

Pathways to Personal Finance

- **7.1** Percents 273
- **7.2** Simple and Compound Interest 277
- **7.3** Buying a Car 282
- **7.4** Buying a House 287

CHAPTER 8

Pathways to Critical Thinking: Sets and Logic

- **8.1** Introduction to Sets 295
- **8.2** Venn Diagrams and Operations with Sets 299
- **8.3** Inductive and Deductive Reasoning 305
- **8.4** Logic Statements, Negations, and Quantified Statements 309
- **8.5** Compound Statements and Connectives 313
- **8.6** Truth Tables 318

CHAPTER 9

Statistical Pathways: Introduction to Probability

- **9.1** Introduction to Probability 325
- **9.2** The Addition Rules of Probability and Odds 330
- **9.3** The Multiplication Rules of Probability and Conditional Probability 335
- **9.4** The Fundamental Counting Principle and Permutations 339
- **9.5** Combinations 343
- **9.6** Using Counting Methods to Find Probability 346

CHAPTER 10
Statistical Pathways: Introduction to Statistics
- 10.1 Collecting Data ... 351
- 10.2 Organizing and Displaying Data ... 355
- 10.3 Measures of Center ... 361
- 10.4 Measures of Dispersion and Percentiles ... 366
- 10.5 The Normal Distribution ... 371

CHAPTER A
Appendix
- A.1 Matrices and Basic Matrix Operations ... 377

Math@Work
- Basic Inventory Management ... 385
- Hospitality Management: Preparing for a Dinner Service ... 387
- Bookkeeper ... 389
- Pediatric Nurse ... 391
- Architecture ... 393
- Statistician: Quality Control ... 395
- Dental Assistant ... 397
- Financial Advisor ... 399
- Market Research Analyst ... 401
- Astronomy ... 403
- Math Education ... 405
- Forensic Scientist ... 407
- Other Careers in Mathematics ... 409

Answer Key ... 411

Math Knowledge Required for Math@Work Career Explorations

The following table summarizes the math knowledge required for each Math@Work career exploration. Use this table to determine when you are ready to explore each career.

Math@Work Career	Whole Numbers	Fractions	Integers	Decimal Numbers	Averages	Percents	Simple Interest	Ratios	Proportions	Geometry	Statistics	Graphing	Linear Equations	Scientific Notation	Greatest Common Factor	Radicals
Basic Inventory Management	✓															
Hospitality Management	✓	✓			✓											
Bookkeeper				✓												
Pediatric Nurse				✓				✓	✓							
Architecture				✓						✓						
Statistician: Quality Control				✓							✓	✓				
Dental Assistant				✓		✓										
Financial Advisor				✓		✓	✓						✓			
Market Research Analyst				✓									✓			
Astronomy				✓										✓		
Math Education			✓												✓	
Forensic Scientist				✓		✓										✓
Other Careers in Mathematics																

🛟 Support

If you have questions or comments we can be contacted as follows:

24/7 Chat: chat.hawkeslearning.com

Phone: (843) 571-2825

E-mail: support@hawkeslearning.com

Web: hawkeslearning.com

Our support hours are 8:00 a.m. to 10:00 p.m. (ET), Monday through Friday.

How to Use the Guided Notebook

There are a variety of elements in this Guided Notebook that will help you on your way to mastering each topic. Here is a rundown of how to use the elements as you work through this notebook.

Fill-in-the-Blanks

1. When there is an incomplete sentence, you will need to write in the ___*missing*___ word(s).

    The ___*missing*___ words can be found by reading through the Learn screens.

Boxed Content

Definitions and **procedures** are highlighted within a box, like the ones shown here. The missing content will vary from box to box. Sometimes an entire definition is missing and sometimes only part of a sentence is missing. Here are two examples of the box variations.

Definition

First term to define: ___*Write the definition here.*___

Second term to define: ___*If there is another term, define it the same way as above.*___

DEFINITION

Terms Related to Probability

___*Outcome*___	An individual result of an experiment.
___*Sample Space*___	The set of all possible outcomes of an experiment.
___*Event*___	Some (or all) of the outcomes from the sample space.

DEFINITION

Properties and **Procedure** boxes are completed in a similar way:

Commutative Property of Multiplication

The order of the numbers in multiplication can be _reversed without changing the product._

For example, _3 · 4 = 12 and 4 · 3 = 12._

PROPERTIES

Subtracting Whole Numbers

1. Write the numbers _vertically_ so that the _place values are lined up in columns._
2. Subtract only the _digits with the same place value._
3. Check by _adding the difference to the subtrahend._ The sum must be _the minuend._

PROCEDURE

▶ Watch and Work

For each Watch and Work, you will need to watch the corresponding video in Learn mode and follow along while completing the example in the space provided.

Example 5 Multiplying Whole Numbers

Multiply: 12 · 35

Solution

The standard form of multiplication is used here to find the product 12 · 35.

$$\begin{array}{r} \overset{1}{1}2 \\ \times\ 35 \\ \hline 60 \\ 360 \\ \hline 420 \end{array}$$

12 · 5 = 60
12 · 30 = 360
Product

✏️ Now You Try It!

After working along with the example video, work through a similar exercise on your own in the space provided.

Example A Multiplying Whole Numbers

Multiply: 25
         × 42
         ────
         1050

1.1 Exercises

Each section has exercises to offer additional practice problems to help reinforce topics that have been covered. The exercises include Concept Check, Practice, Application, and Writing & Thinking questions. The odd answers can be found in the Answer Key at the back of the book.

Concept Check

True/False. Determine whether each statement is true or false. If a statement is false, explain how it can be changed so the statement will be true. (**Note:** There may be more than one acceptable change.)

1. When the given statement is true, you write "True" for the answer.

   True

Practice

For each set of data, find **a.** the mean, **b.** the median, **c.** the mode (if any), and **d.** the range.

2. *Presidents:* The ages of the first five US presidents of the 20th century on the date of their inaugurations were as follows. (The presidents were Roosevelt, Taft, Wilson, Harding, and Coolidge.)

   42,   51,   56,   55,   51

   a. 51   b. 51   c. 51   d. 14

Applications

Solve.

3. *Grades:* Suppose that you have taken four exams and have one more chemistry exam to take. Each exam has a maximum of 100 points and you must average between 75 and 82 points to receive a passing grade of C. If you have scores of 85, 60, 73, and 76 on the first four exams, what is the minimum score you can make on the fifth exam and receive a grade of C?

   81

Writing & Thinking

4. State how to determine the median of a set of data.

   The first step to finding the median is always to arrange the data in order. Once the data is in order, the median is the number in the middle. If there is an even number of items, average the two middle numbers to find the median.

Strategies for Academic Success

Strategies for Academic Success 🎓
How to Read a Math Textbook

Reading a textbook is very different than reading a book for fun. You have to concentrate more on what you are reading because you will likely be tested on the content. Reading a math textbook requires a different approach than reading literature or history textbooks because the math textbook contains a lot of symbols and formulas in addition to words. Here are some tips to help you successfully read a math textbook.

Don't Skim 📖
When reading math textbooks, look at everything: titles, learning objectives, definitions, formulas, text in the margins, and any text that is highlighted, outlined, or in bold. Also pay close attention to any tables, figures, charts, and graphs.

Minimize Distractions
Reading a math textbook requires much more concentration than a novel by your favorite author, so pick a study environment with few distractions and a time when you are most attentive.

🚩 Start at the Beginning
Don't start in the middle of an assigned section. Math tends to build on previously learned concepts and you may miss an important concept or formula that is crucial to understanding the rest of the material in the section.

Highlight and Annotate
Put your book to good use and don't be afraid to add comments and highlighting. If you don't understand something in the text, reread it a couple of times. If it is still not clear, note the text with a question mark or some other notation so you can ask your instructor about it.

Go through Each Step of Each Example 📋
Make sure you understand each step of an example. If you don't understand something, mark it so you can ask about it in class. Sometimes math textbooks leave out intermediate steps to save space. Try working through the examples on your own, filling in any missing steps.

Take Notes   < *This is important!*
Write down important definitions, symbols or notation, properties, formulas, theorems, and procedures. Review these daily as you do your homework and before taking quizzes and tests. Practice rewriting definitions in your own words so you understand them better.

Notes 9-25-17:

- The opposite of a negative integer is a positive integer.

- To add two integers with the same signs add their absolute values and use their common sign

💻 Use Available Resources
Many textbooks have companion websites to help you understand the content. These resources may contain videos that help explain more complex steps or concepts. Try searching the internet for additional explanations of topics you don't understand.

Read the Material Before Class
Try to read the material from your book before the instructor lectures on it. After the lecture, reread the section again to help you retain the information as you look over your class notes.

Understand the Mathematical Definitions + × =
Many terms used in everyday English have a different meaning when used in mathematics. Some examples include equivalent, similar, average, median, and product. Two equations can be equivalent to one another without being equal. An average can be computed mathematically in several ways. It is important to note these differences in meaning in your notebook along with important definitions and formulas.

Try Reading the Material Aloud
Reading aloud makes you focus on every word in the sentence. Leaving out a word in a sentence or math problem could give it a totally different meaning, so be sure to read the text carefully and reread, if necessary.

Questions

1. Explain how taking notes can help you understand new concepts and skills while reading a math textbook.

2. Think of two more tips for reading a math textbook.

Strategies for Academic Success

Tips for Success in a Math Course

Read Your Textbook/Workbook

One of the most important skills when taking a math class is knowing how to read a math textbook. Reading a section before class and then reading it again afterwards is an important strategy for success in a math course. If you don't have time to read the entire assigned section, you can get an overview by reading the introduction or summary and looking at section objectives, headings, and vocabulary terms.

Take Notes

Take notes in class using a method that works for you. There are many different note-taking strategies, such as the Cornell Method and Concept Mapping. You can try researching these and other methods to see if they might work better than your current note-taking system.

Review

While the information is fresh in your mind, read through your notes as soon as possible after class to make sure they are readable, write down any questions you have, and fill in any gaps. Mark any information that is incomplete so that you can get it from the textbook or your instructor later.

Stay Organized

As you review your notes each day, be sure to label them using categories such as definition, theorem, formula, example, and procedure. Try highlighting each category with a different colored highlighter.

Use Study Aids

Use note cards to help you remember definitions, theorems, formulas, or procedures. Use the front of the card for the vocabulary term, theorem name, formula name, or procedure description. Write the definition, the theorem, the formula, or the procedure on the back of the card, along with a description in your own words.

Practice, Practice, Practice!

Math is like playing a sport. You can't improve your basketball skills if you don't practice—the same is true of math. Math can't be learned by only watching your instructor work through problems; you have to be actively involved in doing the math yourself. Work through the examples in the book, do some practice exercises at the end of the section or chapter, and keep up with homework assignments on a daily basis.

Do Your Homework

When doing homework, always allow plenty of time to finish it before it is due. Check your answers when possible to make sure they are correct. With word or application problems, always review your answer to see if it appears reasonable. Use the estimation techniques that you have learned to determine if your answer makes sense.

Understand, Don't Memorize

Don't try to memorize formulas or theorems without understanding them. Try describing or explaining them in your own words or look for patterns in formulas so you don't have to memorize them. For example, you don't need to memorize every perimeter formula if you understand that perimeter is equal to the sum of the lengths of the sides of the figure.

Study

Plan to study two to three hours outside of class for every hour spent in class. If math is your most difficult subject, then study while you are alert and fresh. Pick a study time when you will have the least interruptions or distractions so that you can concentrate.

Manage Your Time

Don't spend more than 10 to 15 minutes working on a single problem. If you can't figure out the answer, put it aside and work on another one. You may learn something from the next problem that will help you with the one you couldn't do. Mark the problems that you skip so that you can ask your instructor about it during the next class. It may also help to work a similar, but perhaps easier, problem.

Questions

1. Based on your schedule, what are the best times and places for you to study for this class?
2. Describe your method of taking notes. List two ways to improve your method.

Strategies for Academic Success
Tips for Improving Math Test Scores

Preparing for a Math Test

- Avoid cramming right before the test and don't wait until the night before to study. Review your notes and note cards every day in preparation for quizzes and tests.
- If the textbook has a chapter review or practice test after each chapter, work through the problems as practice for the test.
- If the textbook has accompanying software with review problems or practice tests, use it for review.
- Review and rework homework problems, especially the ones that you found difficult.
- If you are having trouble understanding certain concepts or solving any types of problems, schedule a meeting with your instructor or arrange for a tutoring session (if your college offers a tutoring service) well in advance of the next test.

Test-Taking Strategies

- Scan the test as soon as you get it to determine the number of questions, their levels of difficulty, and their point values so you can adequately gauge how much time you will have to spend on each question.
- Start with the questions that seem easiest or that you know how to work immediately. If there are problems with large point values, work them next since they count for a larger portion of your grade.
- Show all steps in your math work. This will make it quicker to check your answers later once you are finished since you will not have to work through all the steps again.
- If you are having difficulty remembering how to work a problem, skip it and come back to it later so that you don't spend all of your time on one problem.

After the Test

- The material learned in most math courses is cumulative, which means any concepts you miss on each test may be needed to understand concepts in future chapters. That's why it is extremely important to review your returned tests and correct any misunderstandings that may hinder your performance on future tests.
- Be sure to correct any work you did wrong on the test so that you know the correct way to do the problem in the future. If you are not sure what you did wrong, get help from a peer who scored well on the test or schedule time with your instructor to go over the test.
- Analyze the test questions to determine if the majority came from your class notes, homework problems, or the textbook. This will give you a better idea of how to spend your time studying for the next test.
- Analyze the errors you made on the test. Were they careless mistakes? Did you run out of time? Did you not understand the material well enough? Were you unsure of which method to use?
- Based on your analysis, determine what you should do differently before the next test and where you should focus your time.

Questions

1. Determine the resources that are available to you to help you prepare for tests, such as instructor office hours, tutoring center hours, and study groups.
2. Discuss two additional test taking strategies.

Strategies for Academic Success 🎓

Practice, Patience, and Persistence!

Have you ever heard the phrase "practice makes perfect"? This saying applies to many things in life. You won't become a concert pianist without many hours of practice. You won't become an NBA basketball star by sitting around and watching basketball on TV. The saying even applies to riding a bike. You can watch all of the videos and read all of the books on riding a bike, but you won't learn how to ride a bike without actually getting on the bike and trying to do it yourself. The same idea applies to math. Math is not a spectator sport.

Math is not learned by sleeping with your math book under your pillow at night and hoping for osmosis (a scientific term implying that math knowledge would move from a place of higher concentration—the math book—to a place of lower concentration—your brain). You also don't learn math by watching your professor do hundreds of math problems while you sit and watch. Math is learned by doing. Not just by doing one or two problems, but by doing many problems. Math is just like a sport in this sense. You become good at it by doing it, not by watching others do it. You can also think of learning math like learning to dance. A famous ballerina doesn't take a dance class or two and then end up dancing the lead in The Nutcracker. It takes years of practice, patience, and persistence to get that part.

Now, we aren't suggesting that you dedicate your life to doing math, but at this point in your education, you've already spent quite a few years studying the subject. You will continue to do math throughout college—and your life. To be able to financially support yourself and your family, you will have to find a job, earn a salary, and invest your money—all of which require some ability to do math. You may not think so right now, but math is one of the more useful subjects you will study.

It's important not only to practice math when taking a math course, but also to be patient and not expect immediate success. Just like a ballerina or NBA basketball star, who didn't become exceptional athletes overnight, it will take some time and patience to develop your math skills. Sure, you will make some mistakes along the way, but learn from those mistakes and move on.

Practice, patience, and persistence are especially important when working through applications or word problems. Most students don't like word problems and, therefore, avoid them. You won't become good at working word problems unless you practice them over and over again. You'll need to be patient when working through word problems in math since they will require more time to work than typical math skills exercises. The process of solving word problems is not a quick one and will take patience and persistence on your part to be successful.

Just as you work your body through physical exercise, you have to work your brain through mental exercise. Math is an excellent subject to provide the mental exercise needed to stimulate your brain. Your brain is flexible and it continues to grow throughout your life span—but only if provided the right stimuli. Studying mathematics and persistently working through tough math problems is one way to promote increased brain function. So, when doing mathematics, remember the 3 P's—Practice, Patience, and Persistence—and the positive effects they will have on your brain!

Questions

1. What is another area (not mentioned here) that requires practice, patience, and persistence to master? Can you think of anything you could master without practice?

2. Can you think of an example in your study of math where practice, patience, and persistence have helped you improve?

Strategies for Academic Success 🎓
Note Taking

Taking notes in class is an important step in understanding new material. While there are several methods for taking notes, every note-taking method can benefit from these general tips.

General Tips

- Write the date and the course name at the top of each page.
- Write the notes in your own words and paraphrase.
- Use abbreviations, such as ft for foot, # for number, def for definition, and RHS for right-hand side.
- Copy all figures or examples that are presented during the lecture.
- Review and rewrite your notes after class. Do this on the same day, if possible.

There are many different methods of note taking and it's always good to explore new methods. A good time to try out new note-taking methods is when you rewrite your class notes. Be sure to try each new method a few times before deciding which works best for you. Presented here are three note-taking methods you can try out. You may even find that a blend of several methods works best for you.

Note-Taking Methods

Outline

An outline consists of several topic headings, each followed by a series of indented bullet points that include subtopics, definitions, examples, and other details.

> Example:
> 1. Ratio
>   a. Comparison of two quantities by division.
>   b. Ratio of a to b
>   i. $\dfrac{a}{b}$
>   ii. $a:b$
>   iii. a to b
>   c. Can be reduced
>   d. Common units can cancel

Split Page

The split page method divides the page vertically into two columns with the left column narrower than the right column. Main topics go in the left column and detailed comments go in the right column. The bottom of the page is reserved for a short summary of the material covered.

> Example:
>
> | Keywords: | Notes: |
> |---|---|
> | Ratios | 1. Comparison of two quantities by division |
> | | 2. $\dfrac{a}{b}$, $a:b$, a to b |
> | | 3. Can reduce |
> | | 4. Common units can cancel |
>
> Summary: Ratios are used to compare quantities and units can cancel.

Mapping

The mapping method is the most visual of the three methods. One common way to create a mapping is to write the main idea or topic in the center and draw lines, from the main idea to smaller ideas or subtopics. Additional branches can be created from the subtopics until all of the key ideas and definitions are included. Using a different color for subtopic can help visually organize the topics.

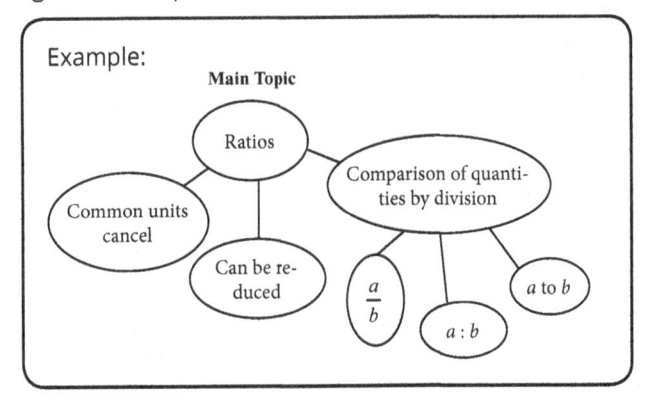

> Example:
>
> Main Topic — Ratios — Comparison of quantities by division — Common units cancel — Can be reduced — $\dfrac{a}{b}$ — $a:b$ — a to b

Questions

1. Find two other note taking methods and describe them.
2. Write five additional abbreviations that you could use while taking notes.

Strategies for Academic Success 🎓
Do I Need a Math Tutor?

If you do not understand the material being presented in class, if you are struggling with completing homework assignments, or if you are doing poorly on tests, then you may need to consider getting a tutor. In college, everyone needs help at some point in time. What's important is to recognize that you need help before it's too late and you end up having to retake the class.

Alternatives to Tutoring

Before getting a tutor, you might consider setting up a meeting with your instructor during their office hours to get help. Unfortunately, you may find that your instructor's office hours don't coincide with your schedule or don't provide enough time for one-on-one help.

Another alternative is to put together a study group of classmates from your math class. Working in groups and explaining your work to others can be very beneficial to your understanding of mathematics. Study groups work best if there are three to six members. Having too many people in a study group may make it difficult to schedule a time for all group members to meet. A large study group may also increase distractions. If you have too few people and those that attend are just as lost as you, then you aren't going to be helpful to each other.

Where to Find a Tutor

Many schools have both group and individual tutoring available. In most cases, the cost of this tutoring is included in tuition costs. If your college offers tutoring through a learning lab or tutoring center, then you should take advantage of it. You may need to complete an application to be considered for tutoring, so be sure to get the necessary paperwork at the start of each semester to increase your chances of getting a tutoring time that works well with your schedule. This is especially important if you know that you struggle with math or haven't taken any math classes in a while.

If you find that you need more help than the tutoring center can provide, or your school doesn't offer tutoring, you can hire a private tutor. The hourly cost to hire a private tutor varies significantly depending on the area you live in along with the education and experience level of the tutor. You might be able to find a tutor by asking your instructor for references or by asking friends who have taken higher-level math classes than you have. You can also try researching the internet for local reputable tutoring organizations in your area.

What to Look for in a Tutor

Whether you obtain a tutor through your college or hire a personal tutor, look for someone who has experience, educational qualifications, and who is friendly and easy to work with. If you find that the tutor's personality or learning style isn't similar to yours, then you should look for a different tutor that matches your style. It may take some effort to find a tutor who works well with you.

How to Prepare for a Tutoring Session

To get the most out of your tutoring session, come prepared by bringing your text, class notes, and any homework or questions you need help with. If you know ahead of time what you will be working on, communicate this to the tutor so they can also come prepared. You should attempt the homework prior to the session and write notes or questions for the tutor. Do not use the tutor to do your homework for you. The tutor will explain to you how to do the work and let you work some problems on your own while he or she observes. Ask the tutor to explain the steps aloud while working through a problem. Be sure to do the same so that the tutor can correct any mistakes in your reasoning. Take notes during your tutoring session and ask the tutor if he or she has any additional resources such as websites, videos, or handouts that may help you.

Questions

1. It's important to find a tutor whose learning style is similar to yours. What are some ways that learning styles can be different?
2. What sort of tutoring services does your school offer?

Strategies for Academic Success

Tips for Improving Your Memory

Experts believe that there are three ways that we store memories: first in the sensory stage, then in short term memory, and finally in long term memory.[1] Because we can't retain all the information that bombards us daily, the different stages of memory act as a filter. Your sensory memory lasts only a fraction of a second and holds your perception of a visual image, a sound, or a touch. The sensation then moves to your short term memory, which has the limited capacity to hold about seven items for no more than 20 to 30 seconds at a time. Important information is gradually transferred to long term memory. The more the information is repeated or used, the greater the chance that it will end up in long term memory. Unlike sensory and short term memory, long term memory can store unlimited amounts of information indefinitely. Here are some tips to improve your chances of moving important information to long-term memory.

Be attentive and focused on the information.

Study in a location that is free of distractions and avoid watching TV or listening to music with lyrics while studying.

Recite information aloud.

Ask yourself questions about the material to see if you can recall important facts and details. Pretend you are teaching or explaining the material to someone else. This will help you put the information into your own words.

Associate the information with something you already know.

Think about how you can make the information personally meaningful—how does it relate to your life, your experiences, and your current knowledge? If you can link new information to memories already stored, you create "mental hooks" that help you recall the information. For example, when trying to remember the formula for slope using rise and run, remember that rise would come alphabetically before run, so rise will be in the numerator in the slope fraction and run will be in the denominator.

Use visual images like diagrams, charts, and pictures.

You can make your own pictures and diagrams to help you recall important definitions, theorems, or concepts.

Split larger pieces of information into smaller "chunks."

This is useful when remembering strings of numbers, such as social security numbers and telephone numbers. Instead of remembering a sequence of digits such as 555777213 you can break it into chunks such as 555 777 213.

Group long lists of information into categories that make sense.

For example, instead of remembering all the properties of real numbers individually, try grouping them into shorter lists by operation, such as addition and multiplication.

Use mnemonics or memory techniques to help remember important concepts and facts.

A mnemonic that is commonly used to remember the order of operations is "Please Excuse My Dear Aunt Sally," which uses the first letter of the words Parentheses, Exponents, Multiplication, Division, Addition, and Subtraction to help you remember the correct order to perform basic arithmetic calculations. To make the mnemonic more personal and possibly more memorable, make up one of your own.

Use acronyms to help remember important concepts or procedures.

An acronym is a type of mnemonic device which is a word made up by taking the first letter from each word that you want to remember and making a new word from the letters. For example, the word HOMES is often used to remember the five Great Lakes in North America where each letter in the word represents the first letter of one of the lakes: Huron, Ontario, Michigan, Erie, and Superior.

Questions

1. Create an original mnemonic or acronym for any math topic covered so far in this course.
2. Explain two ways you can incorporate these tips into your study routine.

---

1 Source: http://science.howstuffworks.com/life/inside-the-mind/human-brain/human-memory2.htm

Strategies for Academic Success 🎓

Overcoming Anxiety

People who are anxious about math are often just not good at taking math tests. If you understand the math you are learning but don't do well on math tests, you may be in the same situation. If there are other subject areas in which you also perform poorly on tests, then you may be experiencing test anxiety.

How to Reduce Math Anxiety

- Learn effective math study skills. Sit near the front of your class and take notes. Ask questions when you don't understand the material. Review your notes after class and read new material before it's covered in class. Keep up with your assignments and do a lot of practice problems.

- Don't accept negative self talk such as "I am not good at math" or "I just don't get it and never will." Maintain a positive attitude and set small math achievement goals to keep you positively moving toward bigger goals.

- Visualize yourself doing well in math, whether it's on a quiz or test, or passing a math class. Rehearse how you will feel and perform on an upcoming math test. It may also help to visualize how you will celebrate your success after doing well on the test.

- Form a math study group. Working with others may help you feel more relaxed about math in general and you may find that other people have the same fears.

- If you panic or freeze during a math test, try to work around the panic by finding something on the math test that you can do. Once you gain confidence, work through other problems you know how to do. Then, try completing the harder problems, knowing that you have a large part of the test completed already.

- If you have trouble remembering important concepts during tests, do what is called a "brain drain" and write down all the formulas and important facts that you have studied on your test or scratch paper as soon as you are given the test. Do this before you look at any questions on the test. Having this information available to you should help boost your confidence and reduce your anxiety. Doing practice brain drains while studying can help you remember the concepts when the test time comes.

How to Reduce Test Anxiety

- Be prepared. Knowing you have prepared well will make you more confident and less anxious.

- Get plenty of sleep the night before a big test and be sure to eat nutritious meals on the day of the test. It's helpful to exercise regularly and establish a set routine for test days. For example, your routine might include eating your favorite food, putting on your lucky shirt, and packing a special treat for after the test.

- Talk to your instructor about your anxiety. Your instructor may be able to make accommodations for you when taking tests that may make you feel more relaxed, such as extra time or a more calming testing place.

- Learn how to manage your anxiety by taking deep, slow breaths and thinking about places or people who make you happy and peaceful.

- When you receive a low score on a test, take time to analyze the reasons why you performed poorly. Did you prepare enough? Did you study the right material? Did you get enough rest the night before? Resolve to change those things that may have negatively affected your performance in the past before the next test.

- Learn effective test taking strategies. See the study skill on Tips for Improving Math Test Scores.

Questions

1. Describe your routine for test days. Think of two ways you can improve your routine to reduce stress and anxiety.

2. Research and describe the accommodations that your instructor or school can provide for test taking.

Strategies for Academic Success

Online Resources

With the invention of the internet, there are numerous resources available to students who need help with mathematics. Here are some quality online resources that we recommend.

HawkesTV

tv.hawkeslearning.com

If you are looking for instructional videos on a particular topic, then start with HawkesTV. There are hundreds of videos that can be found by looking under a particular math subject area such as introductory algebra, precalculus, or statistics. You can also find videos on study skills.

YouTube

www.youtube.com

You can also find math instructional videos on YouTube, but you have to search for videos by topic or key words. You may have to use various combinations of key words to find the particular topic you are looking for. Keep in mind that the quality of the videos varies considerably depending on who produces them.

Google Hangouts

plus.google.com/hangouts

You can organize a virtual study group of up to 10 people using Google Hangouts. This is a terrific tool when schedules are hectic and it avoids everyone having to travel to a central location. You do have to set up a Google+ profile to use Hangouts. In addition to video chat, the group members can share documents using Google Docs. This is a great tool for group projects!

Wolfram|Alpha

www.wolframalpha.com

Wolfram|Alpha is a computational knowledge engine developed by Wolfram Research that answers questions posed to it by computing the answer from "curated data." Typical search engines search all of the data on the Internet based on the key words given and then provide a list of documents or web pages that might contain relevant information. The data used by Wolfram|Alpha is said to be "curated" because someone has to verify its integrity before it can be added to the database, therefore ensuring that the data is of high quality. Users can submit questions and request calculations or graphs by typing their request into a text field. Wolfram|Alpha then computes the answers and related graphics from data gathered from both academic and commercial websites such as the CIA's World Factbook, the United States Geological Survey, financial data from Dow Jones, etc. Wolfram|Alpha uses the basic features of Mathematica, which is a computational toolkit designed earlier by Wolfram Research that includes computer algebra, symbol and number computation, graphics, and statistical capabilities.

Questions

1. Describe a situation where you think Wolfram|Alpha might be more helpful than YouTube, and vice versa.
2. What are some pros and cons to using Google Hangouts?

Strategies for Academic Success 🎓

Preparing for a Final Math Exam

Since math concepts build on one another, a final exam in math is not one you can study for in a night or even a day or two. To pull all the concepts together for the semester, you should plan to start one or two weeks ahead of time. Being comfortable with the material is key to going into the exam with confidence and lowering your anxiety.

Before You Start Preparing for the Exam

1. What is the date, time, and location of the exam? Check your syllabus for the final exam time and location. If it's not on your syllabus, your instructor should announce this information in class.
2. Is there a time limit on the exam? If you experience test anxiety on timed tests, be sure to speak to your professor about it and see if you can receive accommodations that will help reduce your anxiety, such as extended time or an alternate testing location.
3. Will you be able to use a formula sheet, calculator, and/or scrap paper on the exam? If you are not allowed to use a formula sheet, you should write down important formulas and memorize them. Most of the time, math professors will advise you of the formulas you need to know for an exam. If you cannot use a calculator on the exam, be sure to practice doing calculations by hand when you are preparing for the exam and go back and check them using the calculator.

A Week Before the Exam

1. Decide where to study for the exam and with whom. Make sure it's a comfortable study environment with few outside distractions. If you are studying with others, make sure the group is small and that the people in the group are motivated to study and do well on the exam. Plan to have snacks and water with you for energy and to avoid having to delay studying to go get something to eat or drink. Be sure and take small breaks every hour or two to keep focused and minimize frustration.
2. Organize your class notes and any flash cards with vocabulary, formulas, and theorems. If you haven't used flash cards for vocabulary, go back through your notes and highlight the vocabulary. Create a formula sheet to use on the exam, if the professor allows. If not, then you can use the formula sheet to memorize the formulas that will be on the exam.

3. Start studying for the exam. Studying a week before the exam gives you time to ask your instructor questions as you go over the material. Don't spend a lot of time reviewing material you already know. Go over the most difficult material or material that you don't understand so you can ask questions about it. Be sure to review old exams and work through any questions you missed.

3 Days Before the Exam

1. Make yourself a practice test consisting of the problem types. Don't necessarily put the questions in the order that the professor covered them in class.
2. Ask your instructor or classmates any questions that you have about the practice test so that you have time to go back and review the material you are having difficulty with.

The Night Before the Exam

1. Make sure you have all the supplies you will need to take the exam: formula sheet and calculator, if allowed, scratch paper, plain and colored pencils, highlighter, erasers, graph paper, extra batteries, etc.
2. If you won't be allowed to use your formula sheet, review it to make sure you know all the formulas. Right before going to bed, review your notes and study materials, but do not stay up all night to "cram."
3. Go to bed early and get a good night's sleep. You will do better if you are rested and alert.

The Day of the Exam

1. Get up with plenty of time to get to your exam without rushing. Eat a good breakfast and don't drink too much caffeine, which can make you anxious.
2. Review your notes, flash cards, and formula sheet again, if you have time.
3. Get to class early so you can be organized and mentally prepared.

Checklist for the Exam

Date of the Exam: _____   Time of the Exam: _____

Location of the Exam: _____

Items to bring to the exam:

___ calculator and extra batteries   ___ pencils

___ formula sheet   ___ eraser

___ scratch paper   ___ colored pencils or highlighter

___ graph paper   ___ ruler or straightedge

Notes or other things to remember for exam day:

During the Exam

1. Put your name at the top of your exam immediately. If you are not allowed to use a formula sheet, before you even look at the exam, do what is called a "brain drain" or "data dump." Recall as much of the information on your formula sheet as you possibly can and write it either on the scratch paper or in the exam margins if scratch paper is not allowed. You have now transferred over everything on your "mental cheat sheet" to the exam to help yourself as you work through the exam.

2. Read the directions carefully as you go through the exam and make sure you have answered the questions being asked. Also, check your solutions as you go. If you do any work on scratch paper, write down the number of the problem on the paper and highlight or circle your answer. This will save you time when you review the exam. The instructor may also give you partial credit for showing your work. (Don't forget to attach your scratch work to your exam when you turn it in.)

3. Skim the questions on the exam, marking the ones you know how to do immediately. These are the problems you will do first. Also note any questions that have a higher point value. You should try to work these next or be sure to leave yourself plenty of time to do them later.

4. If you get to a problem you don't know how to do, skip it and come back after you finish all the ones you know how to do. A problem you do later may jog your memory on how to do the problem you skipped.

5. For multiple choice questions, be sure to work the problem first before looking at the answer choices. If your answer is not one of the choices, then review your math work. You can also try starting with the answer choices and working backwards to see if any of them work in the problem. If this doesn't work, see if you can eliminate any of the answer choices and make an educated guess from the remaining ones. Mark the problem to come back to later when you review the exam.

6. Once you have an answer for all the problems, review the entire exam. Try working the problems differently and comparing the results or substituting the answers into the equation to verify they are correct. Do not worry about finishing early. You are in control of your own time—and your own success!

Questions

1. Does your syllabus provide any of the information needed for the checklist?

2. Are there any tips or suggestions mentioned here that you haven't thought of before?

Strategies for Academic Success 🎓
Managing Your Time Effectively

Have you ever made it to the end of a day and wondered where all of your time went? Sometimes it feels like there aren't enough hours in the day. Managing your time is important because you can never get that time back. Once it's gone, you have to rush and cram the work into your schedule. Not only will you start feeling stressed out, but you may also find yourself turning in late or incomplete work.

Here are three strategies for managing your time more effectively.

⏱ Time Budgets

Time budgets help you find the time you need to complete necessary projects and tasks. Just like a financial budget shows you how you spend your money, a time budget shows you how you spend your time. You can then identify "wasted" time that could be used more productively.

To begin budgeting your time, assess how much time each week you spend on different types of activities, like Sleep, Meals, Work, Class, Study, Extracurricular, Exercise, Personal, Other, etc.

- What are some activities you'd like to spend more time doing in the future?
- What are some activities you should spend less time doing in the future?

Based on your answers to the questions above, create a weekly time budget. One week contains only 168 hours. If you want to spend more time on a particular activity, you'll need to find that time somewhere. Use a planner to schedule specific blocks of time for study sessions, meals, travel times, and morning/evening routines. As a general rule, you should set aside at least two hours of study time for every one hour of class time. That means that a three-credit course would require at least six hours of outside work per week.

⚖ Breaks

When you are working on an important project or studying for a big exam, you can feel tempted to go as long as possible without taking a break. While staying focused is important, working yourself until you're mentally drained will lower the quality of your work and force you to take even more time recovering.

Just like taking breaks helps your physical body recover, it will also help your brain re-energize and refocus. During study sessions, you should plan to take a break at least once an hour. Study and work breaks should usually last around five minutes. The longer the break, the harder it is to start working again. Some courses have a built-in break during the middle of the class period. Stand up and move around, even if you don't feel tired. Even this little bit of physical movement can help you think more clearly.

📋 Avoiding Multitasking

Multitasking is working on more than one task at a time. When you have several assignments that need to be completed, you may be tempted to save time by working on two or three of them at once. While this strategy might seem like a time-saver, you will probably end up using more time than if you had done each task individually. Not only will you have to switch your focus from one task to the next, but you will also make more mistakes that will need to be corrected later. Multitasking usually ends up wasting time instead of saving it.

Instead of trying to do two things at once, schedule yourself time to work on one task at a time. To-do lists can be helpful tools for keeping yourself focused on finishing one item before moving on to another. You'll do better work and save yourself time.

Questions

1. Are there any areas in your day that are taking up too much of your time, making it hard to devote enough time to more important things?

2. Can you think of a time when multitasking has resulted in lower quality outcome in your experience?

Chapter R
Review of Foundational Math Skills

R.1 Exponents, Prime Numbers, and LCM

R.2 Fractions (Multiplication and Division)

R.3 Fractions (Addition and Subtraction)

R.4 Decimal Numbers

R.5 Bar Graphs, Pictographs, Circle Graphs, and Line Graphs

R.1 Exponents, Prime Numbers, and LCM

Whole Numbers

The **whole numbers** are _____ and the number ____.

    Natural numbers = ____ = _____

    Whole numbers = ____ = _____

Note that 0 is a _____ but not a _____.

With the operation of addition, the numbers being added are called _____ and the result is called the _____.

With the operation of multiplication, the numbers being multiplied are called _____ and the result is called the _____.

Variable

A **variable** is a symbol (generally a letter of the alphabet) that is used to _____.

Exponent and Base

A whole number n is an _____ if it is used to tell _____. The repeated factor a is called the _____. Symbolically,

_____.

Prime Numbers

A **prime number** is a counting number greater than 1 that has _____.

Composite Numbers

A **composite number** is a counting number with
_____ .

Even and Odd Whole Numbers

If a whole number is _____ , it is **even**.

If a whole number is not divisible by 2, it is _____ .

To Find the Prime Factorization of a Composite Number

1. _____ .

2. _____ .

3. _____ .

The **prime factorization** is _____ .

Least Common Multiple (LCM)

The **least common multiple (LCM)** of two (or more) counting numbers is
_____ .

To Find the LCM of a Set of Counting Numbers

1. _____ .

2. _____ .

3. _____ .

▶ Watch and Work

Watch the video for the example in the software and follow along in the space provided.

Example 7: Finding the Least Common Multiple (LCM)

Find the LCM of 8, 10, and 30.

Solution

✏ Now You Try It!

Use the space provided to work out the solution to the next example.

Example A: Finding the Least Common Multiple (LCM)

Find the LCM of 18, 21, and 42.

Solution

Tests for Divisibility

A number is divisible

By 2: _____

By 3: _____

By 4: _____

By 5: _____

By 6: _____

By 9: _____

By 10: _____

R.1 Exercises

Concept Check

True/False. Determine whether each statement is true or false. If a statement is false, explain how it can be changed so the statement is true. (Note: There may be more than one acceptable change.)

1. Nine squared is equal to eighteen.

2. $2^7 = 128$

3. A prime number has exactly 1 factor.

4. All the factors of 30 are 1, 2, 3, 5, 6, 10, 15 and 30.

Practice

5. In the following expression, identify the base and the exponent.

   2^9

6. Use the definition of an exponent to write the given product in exponential form.

$9 \cdot 9 \cdot 9 \cdot 9 \cdot 9$

7. Evaluate the expression.

2^2

8. Evaluate the expression.

12^3

9. Determine whether the following number is prime or composite by dividing by prime numbers. If the number is composite, find at least two pairs of factors whose product is the number.

34

10. Determine whether the following number is prime or composite by dividing by prime numbers. If the number is composite, find at least two pairs of factors whose product is the number.

17

11. Find the prime factorization of the following number.

105

12. Find the prime factorization of the following number.

54

13. Find the LCM of the given set of counting numbers.

12, 40

14. Find the LCM of the given set of counting numbers.

2, 4, 15

Applications

Solve.

15. Two cruise ships leave Charleston at the same time. They take 24 and 15 days, respectively, to reach their destination and return to Charleston. The cruise ships each take continuous trips to and from Charleston. How many days will pass before the two cruise ships leave Charleston on the same day again?

Writing & Thinking

16. Are all prime numbers also odd numbers? Explain your answer.

R.2 Fractions (Multiplication and Division)

Fraction

A **fraction** is a number that can be written in the form

_____ / _____ .

$$\frac{a}{b} \leftarrow \frac{\text{numerator}}{\text{denominator}}$$

Proper Fractions and Improper Fractions

A **proper fraction** is a fraction in which _____. (Proper fractions have values _____.)

An **improper fraction** is a fraction in which _____ _____. (Improper fractions have values _____.)

To Multiply Fractions

1. _____.

2. _____.

$$\frac{a}{b} \cdot \frac{c}{d} = \frac{}{} \quad \text{where } b, d \neq 0.$$

The number 1 is called the _____ for whole numbers; that is, _____ _____ for any whole number a. The number 1 is also the multiplicative identity for fractions since

_____.

If two fractions are equal, they are called _____.

A fraction is reduced to lowest terms _____

_____.

▶ Watch and Work

Watch the video for the example in the software and follow along in the space provided.

Example 9: Multiplying and Reducing Fractions

Multiply and reduce to lowest terms.

$$\frac{18}{35} \cdot \frac{21}{12}$$

Solution

✏ Now You Try It!

Use the space provided to work out the solution to the next example.

Example A: Multiplying and Reducing Fractions

Multiply and reduce to lowest terms.

$$\frac{4}{15} \cdot \frac{35}{48}$$

Solution

Reciprocal

The **reciprocal** of $\frac{a}{b}$ is _____ (where $a \neq 0$ and $b \neq 0$) and _____.

Division

To divide by any nonzero number, _____.

In general,

_____.

R.2 Exercises

Concept Check

True/False. Determine whether each statement is true or false. If a statement is false, explain how it can be changed so the statement is true. (**Note:** There may be more than one acceptable change.)

1. In $\frac{11}{13}$, the denominator is 11.

2. $\frac{17}{0}$ is undefined.

3. To find $\frac{1}{2}$ of $\frac{2}{9}$ requires multiplication.

4. The statement $\frac{1}{3} \cdot \frac{2}{5} = \frac{2}{5} \cdot \frac{1}{3}$ is an example of the associative property of multiplication.

Practice

5. Raise the following fraction to higher terms as indicated.

$$\frac{1}{7} = \frac{?}{35}$$

6. Raise the following fraction to higher terms as indicated.

$$\frac{3}{5} = \frac{?}{60}$$

7. Reduce the following fraction to lowest terms.

   $\dfrac{5}{23}$

8. Reduce the following fraction to lowest terms.

   $\dfrac{29}{39}$

9. Find the product in lowest terms.

   $\dfrac{7}{3} \cdot \dfrac{1}{8}$

10. Find the product in lowest terms.

    $\dfrac{7}{15} \cdot \dfrac{3}{8}$

11. Find the following quotient. Reduce to lowest terms.

    $\dfrac{48}{7} \div \dfrac{48}{27}$

12. Find the following quotient. Reduce to lowest terms.

    $\dfrac{11}{10} \div \dfrac{27}{26}$

Applications

Solve.

13. A bus is carrying 90 passengers, which is $\frac{9}{10}$ of the capacity of the bus. What is the capacity of the bus?

14. There are 3000 students at Canyon High School and $\frac{1}{4}$ of these students are seniors. If $\frac{3}{5}$ of the seniors are opposed to the school forming a rock climbing team and $\frac{9}{10}$ of the remaining students (not seniors) are also opposed to forming a rock climbing team, how many students are in favor of this idea?

Writing & Thinking

15. If two fractions are between 0 and 1, can their product be more than 1? Explain.

R.3 Fractions (Addition and Subtraction)

To Add Fractions with the Same Denominator

1. _____ .

2. _____ .

3. _____ .

To Add Fractions with Different Denominators

1. _____ .

2. _____ .

3. _____ .

4. _____ .

▶ Watch and Work

Watch the video for the example in the software and follow along in the space provided.

Example 3: Adding Fractions with Different Denominators

Find the sum: $\dfrac{3}{8} + \dfrac{13}{12}$

Solution

✏️ Now You Try It!

Use the space provided to work out the solution to the next example.

Example A: Adding Fractions with Different Denominators

Find the sum: $\frac{1}{6} + \frac{3}{10}$

Solution

To Subtract Fractions with the Same Denominator

1. _____ .

2. _____ .

3. _____ .

To Subtract Fractions with Different Denominators

1. _____ .

2. _____ .

3. _____ .

4. _____ .

R.3 Exercises

Concept Check

True/False. Determine whether each statement is true or false. If a statement is false, explain how it can be changed so the statement will be true. (Note: There may be more than one acceptable change.)

1. The final step in adding fractions is to reduce, if possible.

2. The process for finding the LCD is the same as the process for finding the LCM.

3. When subtracting fractions, simply subtract the numerators and the denominators.

4. Subtraction of fractions requires that the fractions have the same denominators.

Practice

5. Compute the sum indicated and simplify your answer.

$$\frac{5}{32} + \frac{15}{32}$$

6. Compute the sum indicated and simplify your answer.

$$\frac{11}{21} + \frac{5}{35}$$

7. Compute the sum indicated and simplify your answer.

$$\frac{6}{18} + \frac{2}{6}$$

8. Compute the difference indicated and simplify your answer.

$$\frac{8}{11} - \frac{4}{11}$$

9. Compute the difference indicated and simplify your answer.

$$\frac{11}{18} - \frac{2}{9}$$

10. Compute the difference indicated and simplify your answer.

$$\frac{17}{20} - \frac{9}{15}$$

Applications

Solve.

11. If your income is $4820 a month and you plan to budget $\frac{3}{4}$ of your income for rent and $\frac{1}{20}$ of your income for food, how much do you plan to spend each month on these two items? Simplify your answer.

12. Three letters weigh $\frac{1}{2}$ ounces, $\frac{1}{3}$ ounces, and $\frac{5}{6}$ ounces. What is the total weight of the letters? Simplify your answer.

Writing & Thinking

13. Explain the steps to follow when adding or subtracting fractions with unlike denominators

R.4 Decimal Numbers

To Read or Write a Decimal Number

1. Read (or write) the _____.

2. Read (or write) the _____.

3. Read (or write) the _____. Then, name the fraction part with the name of the place of the last digit on the _____.

To Compare Two Decimal Numbers

1. Moving **left to right**, compare digits _____.

2. When one compared digit is larger, then the _____.

To Add Decimal Numbers

1. Write the numbers _____.

2. Keep the _____.

3. _____.

4. _____, keeping the decimal point in the sum aligned with the other decimal points.

▶ Watch and Work

Watch the video for the example in the software and follow along in the space provided.

Example 6: Adding Decimal Numbers

Find the sum: $17 + 4.88 + 50.033 + 0.6$

Solution

✏️ Now You Try It!

Use the space provided to work out the solution to the next example.

Example A: Adding Decimal Numbers

Find the sum: $23.8 + 4.2567 + 11 + 3.01$

Solution

To Subtract Decimal Numbers

1. Write the numbers _____ .

2. Keep the _____ .

3. Keep digits with the _____ .

4. _____ , keeping the decimal point in the difference aligned with the other decimal points.

To Multiply Decimal Numbers

1. Multiply the two _____ .

2. Count the total number of places to the
   _____ .

3. Place the decimal point in the product so that the number of decimal places to the
   _____ .

To Divide Decimal Numbers

1. Move the decimal point in the
   _____ .

2. Move the decimal point in the
   _____ .

3. Place the decimal point in the
   _____ .

4. Divide just as with whole numbers. (0s may be added as needed to the dividend.)

Rules for Rounding Decimal Numbers

1. Look at the digit to the right of the place of desired accuracy.

   a. If this digit is less than 5, _____
   _____ and replace all digits to the right with zeros. All digits to the
   _____ .

   b. If this digit is 5 or greater, _____
   _____ and replace all digits to the right with zeros. All digits to the
   _____ . Then, the 9 is
   replaced by 0 and the _____ .

2. Zeros **to the right of the place of accuracy** that are also to the right of the
   _____ .

To Change a Decimal Number to a Percent

1. Move the decimal point two places to the _____ .

2. Write the _____ .

To Change a Percent to a Decimal Number

1. Move the decimal point two places to the _____ .

2. Delete the _____ .

R.4 Exercises

Concept Check

True/False. Determine whether each statement is true or false. If a statement is false, explain how it can be changed so the statement will be true. (Note: There may be more than one acceptable change.)

1. On a number line, any number to the left of another number is larger than that other number.

2. It is important to align the decimal points vertically when adding decimal numbers.

3. The decimal points should be aligned vertically when multiplying decimal numbers.

4. The first step in division with decimal numbers is to move the decimal point in the divisor to the right so that the divisor is a whole number.

Practice

5. Write the following mixed number in decimal notation.

$82\frac{72}{1000}$

6. Write the following number in decimal notation.

Fifty-five hundredths

7. Write the following decimal number in words.

187.942

8. Find the sum.

$42.3 + 74.41 + 173.929$

9. Find the difference.

179.02
-89.2

10. Find the product.

7.6×0.75

11. Find the indicated quotient to the nearest tenth.

$9.69 \div 4.5$

12. Change the following decimal to a percent. Write your answer in percent form.

0.79

13. Change the percent to a decimal.

69.674%

Applications

Solve.

14. Maria wants to buy a new truck for $20,500. She has talked to the loan officer at her credit union and knows that they will loan her $14,000. She must also pay $356.49 for a license fee and $1187.26 for taxes. What amount of cash will she need to buy the truck?

15. If a car travels 336 miles on 12 gallons of gas, how many miles per gallon does it average (to the nearest tenth)?

Writing & Thinking

16. Why is it important that the decimal points and numbers be aligned vertically when adding or subtracting decimals?

R.5 Bar Graphs, Pictographs, Circle Graphs, and Line Graphs

Four Types of Graphs and Their Purposes

1. _____ : To emphasize comparative amounts

2. _____ : To emphasize the topic being related as well as the the quantities

3. _____ : To help in understanding percents or parts of a whole; also called **pie charts**

4. _____ : To indicate tendencies or trends over a period of time

Properties of Graphs

1. They should be _____ .

2. They should be _____ .

3. They should have _____ .

Watch and Work

Watch the video for the example in the software and follow along in the space provided.

Example 3: Reading a Circle Graph

Examine the following circle graph. This graph shows the percent of a household's annual income they plan to budget for various expenses. Suppose the household has an annual income of $45,000. Use the information in the graph to calculate how much money will be budgeted for each expense.

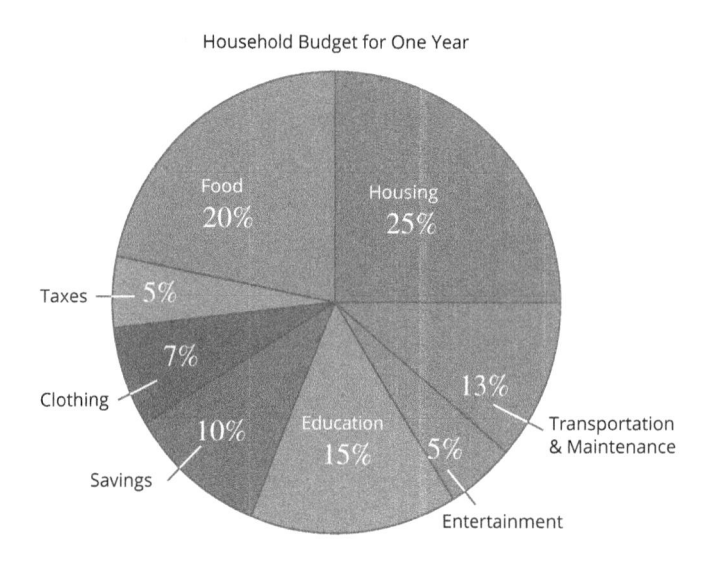

Household Budget for One Year

- Food 20%
- Housing 25%
- Taxes 5%
- Clothing 7%
- Savings 10%
- Education 15%
- Entertainment 5%
- Transportation & Maintenance 13%

Solution

Item		Amount

R.5 Bar Graphs, Pictographs, Circle Graphs, and Line Graphs

✏️ Now You Try It!

Use the space provided to work out the solution to the next example.

Example A: Reading a Circle Graph

Using the circle graph in Example 2, how much will the family spend on each of the following expenses if the family income increases to $55,000 and the percent budgeted for each expense does not change?

   a. Housing

   b. Savings

   c. Clothing

   d. Food

Solution

R.5 Exercises

Concept Check

True/False. Determine whether each statement is true or false. If a statement is false, explain how it can be changed so the statement will be true. (Note: There may be more than one acceptable change.)

  1. Graphs should always be clearly labeled, easy to read, and have appropriate titles.

  2. Circle graphs show trends over a period of time.

  3. Line graphs are used to indicate a trend over a period of time.

  4. Pictographs are used to illustrate percentages as part of a whole.

Practice

5. The following bar graph shows the per capita personal incomes for six states in 2015. Use this bar graph to answer the questions.

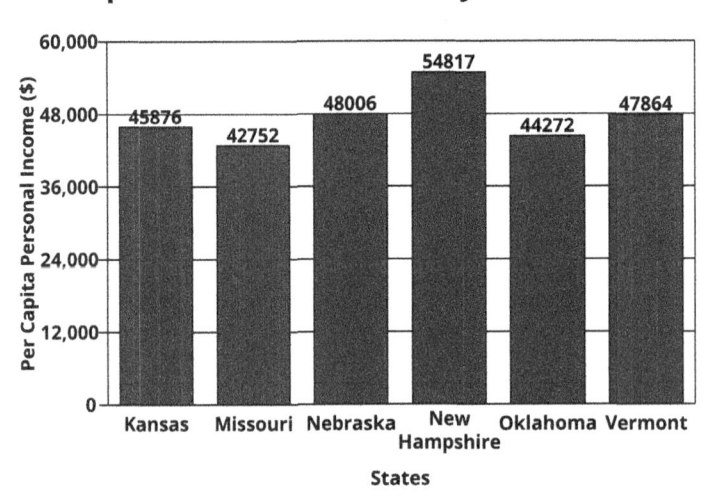

Per Capita Personal Income by State for 2015

State	Per Capita Personal Income ($)
Kansas	45876
Missouri	42752
Nebraska	48006
New Hampshire	54817
Oklahoma	44272
Vermont	47864

    a. Find the lowest per capita personal income for the six states shown. Round your answer to the nearest hundredth, if necessary.

    b. Find the highest per capita personal income for the six states shown. Round your answer to the nearest hundredth, if necessary.

    c. Which state of the six shown has the highest per capita personal income?

    d. What is the difference in per capita personal income between New Hampshire and Missouri? Round your answer to the nearest hundredth, if necessary.

    e. If a single resident of Missouri makes $40,000 in 2015, what percent of the per capita personal income for Missouri was his/her salary? Round your answer to the nearest hundredth of a percent, if necessary.

6. The Pizza Pie 'N Go sells about 1600 one-topping pizzas each month. The circle graph displays the most requested one-topping pizzas, by percentage, for one month.

Most Popular One-Topping Pizzas

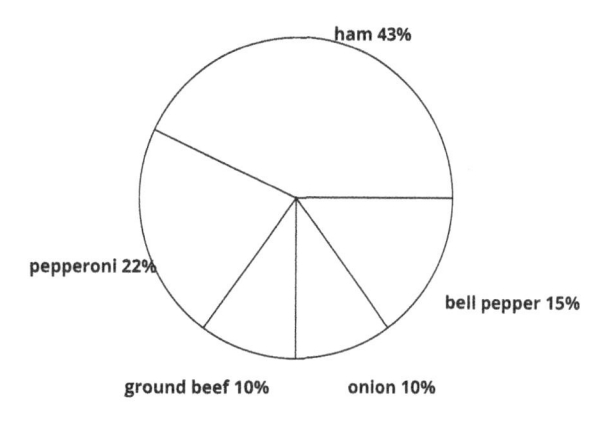

ham 43%
pepperoni 22%
bell pepper 15%
ground beef 10%
onion 10%

a. Find the number of ham pizzas sold each month. Round your answer to the nearest integer.

b. Find the number of pepperoni pizzas sold each month. Round your answer to the nearest integer.

c. Find the number of ground beef pizzas sold each month. Round your answer to the nearest integer.

d. Find the number of onion pizzas sold each month. Round your answer to the nearest integer.

e. Find the number of bell pepper pizzas sold each month. Round your answer to the nearest integer.

7. The following line graph shows the per game scoring averages for Lebron James from the 2013 NBA season to the 2018 NBA season. Use the line graph to answer the questions.

NBA Player Scoring Averages: 2013 - 2018

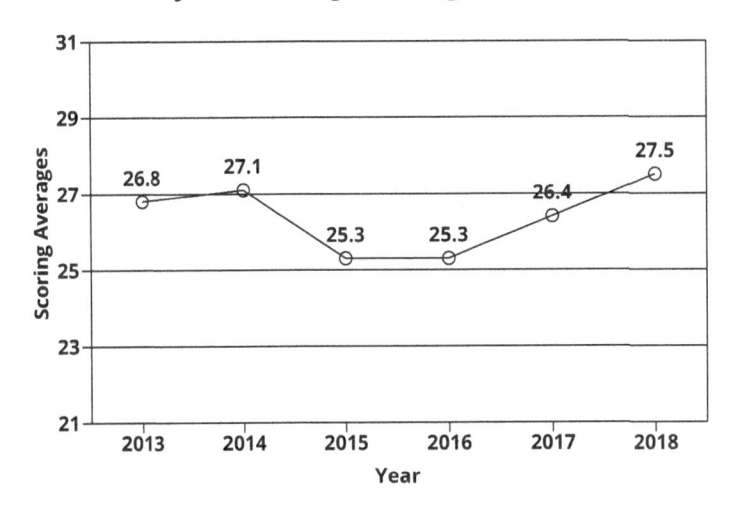

Year	2013	2014	2015	2016	2017	2018
Scoring Average	26.8	27.1	25.3	25.3	26.4	27.5

a. Find the lowest per game scoring average for the six seasons shown.

b. Find the highest per game scoring average for the six season shown.

Writing & Thinking

8. State three properties or characteristics that should be true of all graphs so that they can communicate numerical data quickly and easily.

Chapter 1
Algebraic Pathways: Real Numbers and Algebraic Expressions

1.1 The Real Number Line and Absolute Value

1.2 Operations with Real Numbers

1.3 Problem Solving with Real Numbers

1.4 Square Roots and Order of Operations with Real Numbers

1.5 Properties of Real Numbers

1.6 Simplifying and Evaluating Algebraic Expressions

1.7 Translating English Phrases and Algebraic Expressions

1.1 The Real Number Line and Absolute Value

The set of numbers

$$\mathbb{N} = \{1, 2, 3, 4, 5, 6, 7, 8, 9, 10, 11, \ldots\}$$

is called the _____ or _____. Putting 0 with the set of _____ gives the set of _____.

Integers

The set of numbers consisting of the
_____ :
_____

Variables

A **variable** is a symbol (generally a _____) that is used to
_____.

Rational Numbers

A **rational number** is a number that can be written _____
_____

OR

A **rational number** is a number that can be written in
_____.

Numbers that cannot be written as fractions with integer numerators and denominators are called _____.

Chapter 1 Algebraic Pathways: Real Numbers and Algebraic Expressions

▶ Watch and Work

Watch the video for the example in the software and follow along in the space provided.

Example 4: Graphing Sets of Numbers

Graph the set of **real numbers** $\left\{-\dfrac{3}{4}, 0, 1, 1.5, 3\right\}$.

Solution

✏ Now You Try It!

Use the space provided to work out the solution to the next example.

Example A: Graphing Sets of Numbers

Graph the set of real numbers $\left\{-2.5, -1, 0, \dfrac{5}{4}, 4\right\}$.

Solution

Symbols of Equality and Inequality

$=$ _____

$\neq$ _____

$<$ _____

$>$ _____

$\leq$ _____

$\geq$ _____

1.1 The Real Number Line and Absolute Value   35

Absolute Value

The **absolute value** of a real number is _____. Note that the absolute value of a _____.

1.1 Exercises

Concept Check

True/False. Determine whether each statement is true or false. If a statement is false, explain how it can be changed so the statement is true. (**Note:** There may be more than one acceptable change.)

1. On a number line, smaller numbers are always to the left of larger numbers.

2. The absolute value of a negative number is a positive number.

3. All whole numbers are also integers.

4. Zero is a positive number.

Practice

5. Plot the set $\{-3, -1, 1, 5\}$ on the number line.

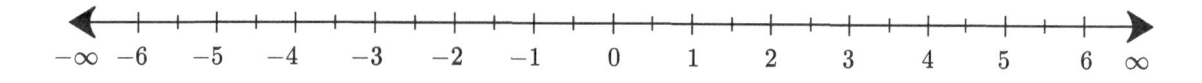

6. Plot the set $\left\{-4, \dfrac{1}{2}, 1, 4\right\}$ on the number line.

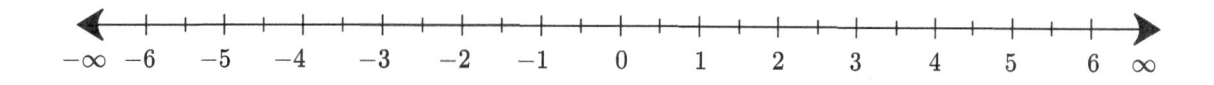

7. Classify the following number.

    1

8. Classify the following number.

    9.3333...

9. Determine whether the following statement is true or false. If it is false, rewrite it in a form that is a true statement.

    $14 > -14$

10. Select the symbol from the set $\{<, >, =\}$ that can be placed in the blank to make a true statement.

    $\dfrac{3}{4} \underline{} \dfrac{5}{2}$

11. Select the symbol from the set $\{<, >, =\}$ that can be placed in the blank to make a true statement.

    $4.2 \underline{} -5.9$

12. Find the real numbers that satisfy the equation below.

    $|x| = 40.5$

13. Find the real numbers that satisfy the equation below.

    $|x| = -4.6$

14. Fill in the blank with the appropriate symbol: $<, >,$ or $=$.

    $-14.9 \underline{} |14.9|$

Applications

Solve.

15. The Alvin is a manned deep-ocean research submersible that has explored the wreck of the Titanic. The operating depth of the Alvin is 4500 meters below sea level.

16. The Mariana trench is the deepest known location on the Earth's ocean floor. The deepest known part of the Mariana Trench is approximately 11 kilometers below sea level.

Writing & Thinking

17. Explain, in your own words, how an expression such as $-y$ might represent a positive number.

1.2 Operations with Real Numbers

Rules for Addition with Real Numbers

1. To add two real numbers with **like signs**,

    a. _____

    b. use the _____ .

2. To add two real numbers with **unlike signs**,

    a. _____

    b. use the _____ .

The **perimeter** of a geometric figure is the _____ .

Rule for Subtraction with Real Numbers

For real numbers a and b,

_____ .

To subtract b, add the _____ of b.

1.2 Operations with Real Numbers   39

▶ Watch and Work

Watch the video for the example in the software and follow along in the space provided.

Example 6: Application: Calculating Change in Value

At noon on Tuesday the temperature was 34 °F. By noon on Thursday the temperature had dropped to −5 °F. How much did the temperature drop between Tuesday and Thursday?

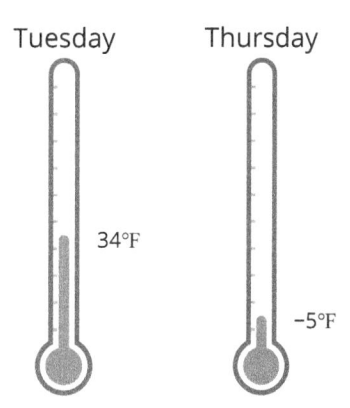

Tuesday   Thursday

34°F

−5°F

Solution

✏ Now You Try It!

Use the space provided to work out the solution to the next example.

Example A: Calculating Change in Value

At 3 p.m. on Friday the temperature was 63 °F. By midnight the temperature had changed to 44 °F. How much did the temperature change between 3 p.m. and midnight?

Solution

Rules for Multiplying Positive and Negative Real Numbers

For positive real numbers a and b,

1. The product of two positive numbers is _____ .

2. The product of two negative numbers is _____ .

3. The product of a positive number and a negative number is
   _____ .

Area of a Rectangle

The **area** of a rectangle (measured in square units) is found by
_____ .

Rules for Dividing Positive and Negative Real Numbers

For positive real numbers a and b,

1. The quotient of two positive numbers is _____ .

2. The quotient of two negative numbers is _____ .

3. The quotient of a positive number and a negative number is
   _____ .

1.2 Exercises

Concept Check

True/False. Determine whether each statement is true or false. If a statement is false, explain how it can be changed so the statement is true. (**Note:** There may be more than one acceptable change.)

1. The sum of a positive number and a negative number is always positive.

2. The sum of two positive numbers can equal zero.

3. The expression "15 − 7" can be thought of as "fifteen plus negative seven."

4. If two numbers have the same sign, both the product and the quotient of the two numbers will be negative.

5. The mean of a set of numbers is always positive.

Practice

6. Compute the value of the following sum.

$-16.6 + 3.7$

7. Compute the value of the following sum.

$-60 + (-79) + 42$

8. Find the perimeter of the rectangle.

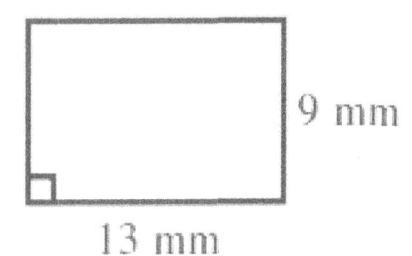

9 mm

13 mm

9. Find the product.

$3 \cdot (-5) \cdot (-4)$

10. Find the product.

$(-33) \cdot 0$

11. Find the quotient.

$$\frac{45}{3}$$

12. Find the quotient.

$$\frac{46}{0}$$

Applications

Solve.

13. A pilot flew a plane from an altitude of 10000 feet to an altitude of 3600 feet. What was the change in altitude?

14. In April, Mr. Burton opened a checking account and made deposits of $968, $1387, $986, and $369. He also wrote checks for $193, $480, $52, $468, and $545. What was his balance at the end of the month?

15. Find the area of the rectangle:

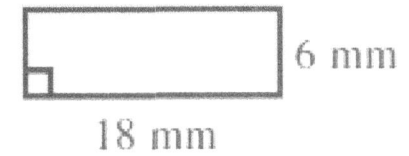

6 mm
18 mm

Writing & Thinking

16. If you multiply an odd number of negative numbers together, do you think that the product will be positive or negative? Explain your reasoning.

1.3 Problem Solving with Real Numbers

Basic Strategy for Solving Word Problems

1. READ: _____ .

2. SET UP: Draw any type of figure or diagram that might be helpful and _____ .

3. SOLVE: _____ .

4. CHECK: Check your work and _____ .

To Find the Average of a Set of Numbers

1. Find the _____ .

2. Divide this sum by the _____ .

▶ Watch and Work

Watch the video for the example in the software and follow along in the space provided.

Example 6: Calculating an Average

Find the average of the following set of numbers: 15, 8, 90, 35, 27.

Solution

✏️ Now You Try It!

Use the space provided to work out the solution to the next example.

Example A: Calculating an Average

Find the average of the following set of numbers: 18, 29, 6, 33, 14, 26.

Solution

1.3 Exercises

Concept Check

True/False. Determine whether each statement is true or false. If a statement is false, explain how it can be changed so the statement is true. (**Note:** There may be more than one acceptable change.)

1. Averages are found by performing addition and then division.

2. The sum of 312 and 4 is 1248.

3. The word "quotient" indicates multiplication.

4. After reading a problem carefully, the next step might be to make a diagram or draw a figure.

Practice

5. Find the average (mean) of the set of numbers.

8, 6, 5, 4, 6, 7, 6

Applications

Solve.

6. Three letters weigh $\frac{1}{2}$ ounces, $\frac{1}{3}$ ounces, and $\frac{5}{6}$ ounces. What is the total weight of the letters? Simplify your answer.

7. There are 3000 students at Canyon High School and $\frac{1}{4}$ of these students are seniors. If $\frac{3}{5}$ of the seniors are in favor of the school forming a rock climbing team and $\frac{7}{10}$ of the remaining students (not seniors) are also in favor of forming a rock climbing team, how many students are opposed to this idea?

8. If a car travels 336 miles on 12 gallons of gas, how many miles per gallon does it average (to the nearest tenth)?

9. A new living room set costs $1210. Not included in the price are the cost of delivery, which is $60, and tax, which is $107. If June pays $448 today, how much more does she still owe to the furniture store?

10. If you opened an account with $1947 and then wrote checks for $189, $77, $528, $131, and $75, what would your balance be?

11. Consider the following word problem:

    The temperature readings for 20 days at 11 AM at a local ski resort were recorded as follows:

    19° 0° −1° −9° −5° 5° −5° 21° 18° −6°

    14° 10° 1° 6° 25° 4° 25° 9° 5° 28°

    What was the average of the recorded temperatures for these 20 days? Round your answer to the nearest tenth.

Writing & Thinking

12. Give an example where you might use average (other than in a class).

1.4 Square Roots and Order of Operations with Real Numbers

A number is **squared** when it is _____. If a whole number is squared, the result is called _____.

The symbol $\sqrt{}$ is called a _____. The number under the radical sign is called the _____. The complete expression, such as $\sqrt{25}$, is called a _____.

> ### Rules for Order of Operations
>
> 1. Simplify within grouping symbols, such as _____
>   _____.
>
> 2. Find any powers indicated _____.
>
> 3. Moving from **left to right**, perform any _____
>   _____.
>
> 4. Moving from **left to right**, perform any _____
>   _____.

▶ Watch and Work

Watch the video for the example in the software and follow along in the space provided.

Example 4: Using the Order of Operations with Real Numbers

Simplify: $(2-5)^2 + |2-5^2| - 2^3$

Solution

✏️ Now You Try It!

Use the space provided to work out the solution to the next example.

Example A: Using the Order of Operations with Real Numbers

Simplify: $(1-3)^2 + |9 - 4^2| - 1^3$

Solution

1.4 Exercises

Concept Check

True/False. Determine whether each statement is true or false. If a statement is false, explain how it can be changed so the statement is true. (**Note:** There may be more than one acceptable change.)

1. In the expression $\sqrt{81}$, the number 9 is the radicand.

2. When following the rules for order of operations, powers indicated by exponents should be evaluated last.

3. The square root symbol is a grouping symbol.

4. A well-known mnemonic device for remembering the rules for order of operations is SADMEP.

Practice

5. Find the value of the indicated square root.

    $\sqrt{1}$

6. Determine whether the indicated number is a perfect square.

    1

1.4 Square Roots and Order of Operations with Real Numbers   49

7. Find the value of the indicated expression.

$13 - 2\sqrt{4}$

8. Find the value of the following expression using the rules for order of operations.

$(-21) \cdot 3 - (4 + 3)^2$

9. Find the value of the following expression using the rules for order of operations.

$6 + [24 \div [-5 - 3 \cdot (-1)]]$

10. Find the value of the following expression using the rules for order of operations.

$\dfrac{1}{4} \div \dfrac{3}{7} - \dfrac{2}{3} \div \dfrac{4}{6}$

Applications

Solve.

11. The Matthews family, a family of 4, is planning a trip to New York City. During their visit, they want to see the Broadway play Matilda. The tickets cost $102 each. The Matthews purchase the tickets online and the website charges a service fee of $7.50 per ticket. The website is running a sale where the Matthews can get 10% off of their entire purchase.

   a. Write an expression to describe how much of a discount the Matthews will receive on their purchase.

   b. What is the final purchase price of the tickets?

12. Dennis overdrew his checking account and ended up with a balance of −$42. The bank charged a $35 overdraft fee and an additional $5 fee for every day the account was overdrawn. Dennis left his account overdrawn for 3 days.

   a. Write an expression to show the balance of Dennis's checking account after 3 days.

b. Simplify the expression in Part a. to find the balance of Dennis's checking account after 3 days.

Writing & Thinking

13. Determine the difference between a radical sign, radicand, and radical expression.

1.5 Properties of Real Numbers

Properties of Addition and Multiplication

In this table a, b, and c are real numbers.

Name of Property	For Addition	For Multiplication
Commutative Property		
Associative Property		
Identity		
Inverse		

Zero-Factor Law

_____

_____

Distributive Property of Multiplication over Addition

_____

▶ Watch and Work

Watch the video for the example in the software and follow along in the space provided.

Example 2: Identifying Properties of Addition and Multiplication

For each of the following equations, state the property illustrated and show that the statement is true for the value given for the variable by substituting the value in the equation and evaluating.

a. $x + 14 = 14 + x$ given that $x = -4$

b. $(3 \cdot 6)x = 3(6x)$ given that $x = 5$

c. $12(y + 3) = 12y + 36$ given that $y = -2$

Solution

✏ Now You Try It!

Use the space provided to work out the solution to the next example.

Example A: Identifying Properties of Addition and Multiplication

State the property illustrated and show that the statement is true for the value given for the variable.

a. $x + 21 = 21 + x$ given that $x = -7$

b. $(5 \cdot 4) x = 5 (4x)$ given that $x = 2$

c. $11 (y + 3) = 11y + 33$ given that $y = -4$

Solution

1.5 Exercises

Concept Check

True/False. Determine whether each statement is true or false. If a statement is false, explain how it can be changed so the statement is true. (**Note:** There may be more than one acceptable change.)

1. Changing the order of the numbers in an addition problem is allowed because of the associative property of addition.

2. The equation $(8 \cdot 2) \cdot 5 = 8 \cdot (2 \cdot 5)$ is an example of the associative property of multiplication.

3. The additive identity of all numbers is 1.

4. The commutative property works for division and subtraction.

Practice

5. Identify the property of real numbers illustrated in the following equation.

$8 + [y^3 \cdot (-2)] = [y^3 \cdot (-2)] + 8$

6. Identify the property of real numbers illustrated in the following equation.

$(-6)[z \cdot (-2)] = [z \cdot (-2)](-6)$

7. Identify the property of real numbers illustrated in the following equation.

$(-5)[(-3) \cdot (-2)] = [(-5) \cdot (-3)](-2)$

8. Identify the property of real numbers illustrated in the following equation.

$2[(-6y^3) + 6z] = -12y^3 + 12z$

9. Identify the property of real numbers illustrated in the following equation.

$(z^2 + 2) + 0 = (z^2 + 2)$

Applications

Solve.

10. Jessica works part-time at a retail store and makes $11 an hour. During one week, she worked $6\frac{1}{2}$ hours on Monday and $4\frac{1}{4}$ hours on Thursday.

   a. Determine the amount of money she earned during the week by evaluating the expression $\$11 \cdot \left(6\frac{1}{2} + 4\frac{1}{4}\right)$.

   b. Rewrite this expression to remove the parentheses using one of the properties talked about in this section.

   c. What property did you use in Part b. to rewrite the expression?

11. Robin went to the grocery store to buy a few items she needed in order to cook dinner. She bought milk for $3.99, rolls for $2.25, a package of steaks for $12.01, and some marinade for $1.75. Before getting to the checkout line, Robin remembered that she only had $20 in her purse. Did she have enough money to buy the food items if the store does not charge sales tax on food?

   a. Write an expression to find the total of Robin's food purchases. Do not simplify.

   b. Robin doesn't have a calculator to determine the total cost of her items. She wants to make sure that she has enough money to buy them. Rearrange the expression from Part a. so that she could quickly find the total using mental math.

   c. What properties did you use in Part b. to rewrite the expression?

   d. Did Robin have enough money to purchase all of the items?

Writing & Thinking

12. a. The distributive property illustrated as $a(b+c) = ab + ac$ is said to "distribute multiplication over addition." Explain, in your own words, the meaning of this phrase.

b. What would an expression that "distributes addition over multiplication" look like? Explain why this would or would not make sense.

1.6 Simplifying and Evaluating Algebraic Expressions

Like Terms

Like terms (or **similar terms**) are terms that are _____ _____.

Combining Like Terms

To **combine like terms,** _____.

To Evaluate an Algebraic Expression

1. _____.

2. _____.

3. Follow the rules for order of operations.

(**Note:** Terms separated by + and − signs may be evaluated _____ _____.)

▶ Watch and Work

Watch the video for the example in the software and follow along in the space provided.

Example 5: Simplifying and Evaluating Algebraic Expressions

Simplify and evaluate $3ab - 4ab + 6a - a$ for $a = 2, b = -1$.

Solution

Now You Try It!

Use the space provided to work out the solution to the next example.

Example A: Simplifying and Evaluating Algebraic Expressions

Simplify and evaluate $5ab - 8ab + 2a - 3a$ for $a = 3$, $b = 1$.

Solution

1.6 Exercises

Concept Check

True/False. Determine whether each statement is true or false. If a statement is false, explain how it can be changed so the statement is true. (**Note:** There may be more than one acceptable change.)

1. A variable that does not appear to have an exponent has an exponent of 1.

2. In the term $-9x$, nine is being subtracted from x.

3. In the term "$12a$," 12 is the constant.

4. Like terms have the same coefficients.

Practice

5. Simplify the algebraic expression by combining like (or similar) terms.

   $0.7x - 2x$

6. Simplify the algebraic expression by combining like (or similar) terms.

   $0.7x - 7 + 6x$

7. Simplify the algebraic expression by combining like (or similar) terms.

$-0.2 - 2x + 0.6 + 7x$

8. Simplify the algebraic expression by performing the indicated operations and combining the similar (or like) terms.

$2(x + 3) - 3(y + 3)$

9. Evaluate the following algebraic expression at $b = 3$ and simplify your answer.

$11b + \dfrac{-3b - 5b}{-8}$

10. First (a) simplify the expression and then (b) evaluate the expression for $x = -5$.

$2x + 4[x - 3(2 + x)]$

Applications

Solve.

11. An apartment management company owns a property with 100 units. The company has determined that the profit made per month from the property can be calculated using the equation $P = -10x^2 + 1500x - 6000$, where x is the number of units rented per month. How much profit does the company make when 80 units are rented?

12. A ball is thrown upward from an initial height of 96 feet with an initial velocity of 16 feet per second. After t seconds, the height of the ball can be described by the expression $-16t^2 + 16t + 96$. What is the height of the ball after 3 seconds?

Writing & Thinking

13. Define constant and variable. Explain why those particular words are used.

1.7 Translating English Phrases and Algebraic Expressions

Key Words To Look For When Translating Phrases				
Addition	**Subtraction**	**Multiplication**	**Division**	**Exponent (Powers)**
_____	_____	_____	_____	_____
_____	_____	_____	_____	_____
	_____	_____	_____	
_____	_____	_____		
_____	_____	_____		
_____	_____			

Division and subtraction are done with the values in the _____ that they are given in the problem.

An **ambiguous phrase** is one whose meaning is

_____ .

▶ Watch and Work

Watch the video for the example in the software and follow along in the space provided.

Example 3: Translating Algebraic Expressions into English Phrases

Change each algebraic expression into an equivalent English phrase. In each case translate the variable as "a number."

	Algebraic Expression	Possible English Phrase
a.	$5x$	_____
b.	$2n + 8$	_____
c.	$3(a - 2)$	_____

Chapter 1 Algebraic Pathways: Real Numbers and Algebraic Expressions

✏️ Now You Try It!

Use the space provided to work out the solution to the next example.

Example A: Translating Algebraic Expressions into English Phrases

Change each algebraic expression into an equivalent English phrase.

a. $10x$

b. $4a + 7$

c. $7(n - 5)$

Solution

1.7 Exercises

Concept Check

True/False. Determine whether each statement is true or false. If a statement is false, explain how it can be changed so the statement is true. (**Note:** There may be more than one acceptable change.)

1. The order in which the values are given is particularly important when working with subtraction and division problems.

2. "More than" and "increased by" are key phrases specifying the operation of subtraction.

3. Division is indicated by the phrase "five less than a number."

4. Key phrases for parentheses can be used to limit ambiguity in English phrases.

Practice

5. Translate the following phrase into an algebraic expression. Do not simplify. Use the variable names "x" or "y" to describe the unknowns.

three added to a number

6. Translate the following phrase into an algebraic expression. Do not simplify.

3 more than the sum of y and x

7. Translate the following phrase into an algebraic expression. Do not simplify.

x multiplied by the product of 2 and y

8. Translate the following phrase into an algebraic expression. Do not simplify. Use the variable names "x" or "y" to describe the unknowns.

the sum of two numbers plus 3

9. Translate the following algebraic expression into an equivalent English phrase.

$3 + z$

10. Translate the following algebraic expression into an equivalent English phrase.

$5(7 - x)$

Writing & Thinking

11. Explain the difference between $5(n + 3)$ and $5n + 3$ when converting from algebra to English.

Chapter 2
Algebraic Pathways: Linear Equations and Inequalities

2.1 Solving One-Step Linear Equations

2.2 Solving Multi-Step Linear Equations

2.3 Working with Formulas

2.4 Applications of Linear Equations

2.5 Ratios, Rates, and Proportions

2.6 Modeling Using Variation

2.7 Solving Linear Inequalities in One Variable

2.1 Solving One-Step Linear Equations

An _____ is a statement that two algebraic expressions are equal. If an equation contains a variable, any number that gives a true statement when substituted for the variable is called a _____ to the equation. The solutions to an equation form _____. The process of finding the solution set is called _____.

Linear Equation in x

If a, b, and c are **constants** and $a \neq 0$, then a **linear equation in x** is _____ _____ _____.

Addition Principle of Equality

If the same algebraic expression is added to _____ _____. Symbolically, if A, B, and C are algebraic expressions, then the equations

_____

and

_____

_____.

Procedure for Solving Linear Equations that Simplify to the Form $x + b = c$

1. Combine _____.

2. Use the **addition principle of equality** and
   _____. The objective is to _____
   _____
   _____.

3. Check your answer by
   _____.

▶ Watch and Work

Watch the video for the example in the software and follow along in the space provided.

Example 2: Solving Linear Equations of the Form $x + b = c$

Solve the equation: $x - 3 = 7$

Solution

✏ Now You Try It!

Use the space provided to work out the solution to the next example.

Example A: Solving Linear Equations of the Form $x + b = c$

Solve the equation: $x - 5 = 12$

Solution

2.1 Solving One-Step Linear Equations

Multiplication (or Division) Principle of Equality

If both sides of an equation are multiplied by (or divided by) the _____

_____ . Symbolically, if A and B are algebraic expressions and C is any nonzero constant, then the equations

Procedure for Solving Linear Equations that Simplify to the Form $ax = c$

1. Combine _____ .

2. Use the **multiplication** (or **division**) **principle of equality** and multiply both sides of the equation by the _____ (**or divide both sides by** _____). The coefficient of the variable will become _____ .

3. Check your answer by _____ .

2.1 Exercises

Concept Check

True/False. Determine whether each statement is true or false. If a statement is false, explain how it can be changed so the statement is true. (**Note:** There may be more than one acceptable change.)

1. When an algebraic expression is added to both sides of an equation, the new equation has the same solutions as the original equation.

2. The process of finding the solution set to an equation is called simplifying the equation.

3. A linear equation in x is also called a first-degree equation in x.

68 Chapter 2 Algebraic Pathways: Linear Equations and Inequalities

4. Equations with the same solutions are said to be equivalent equations.

Practice

5. Determine whether the given number is a solution to the given equation by substituting and then evaluating.

$y + (-5) = -3; y = 2$

6. Determine whether the given number is a solution to the given equation by substituting and then evaluating.

$-1 - |y| = -8; y = -7$

7. Solve the linear equation using equivalent equations to isolate the variable. Express your solution as an integer, as a simplified fraction, or as a decimal number.

$w + 11 = 4$

8. Solve the linear equation using equivalent equations to isolate the variable. Express your solution as an integer, as a simplified fraction, or as a decimal number.

$9.4 - 7.3 = 7.8a - 6.8a - 7.5$

9. Solve the following linear equation using equivalent equations to isolate the variable. Express your answer as an integer, as a simplified fraction, or as a decimal number rounded to two places.

$-6 = 2u$

10. Solve the following linear equation using equivalent equations to isolate the variable. Express your answer as an integer, as a simplified fraction, or as a decimal number rounded to two places.

$-\dfrac{6}{5}y + \dfrac{2}{5}y = \dfrac{2}{3} - \dfrac{1}{3}$

Applications

Solve.

11. An office supply company offers a number of different packages to help save their customers money. One of the packages includes dry erase markers, 22 boxes of paper clips, and 22 boxes of printer ink. If the total number of boxes in the package is 67, use the formula $x + 2(22) = 67$ to find the number of boxes of dry erase markers that come in the package.

12. Janice lives 154 miles away from her cousin. The distance between Janice's house and her cousin's house is 9 times farther than the distance between Janice's house and her best friend's house. Solve the equation $9x = 154$ to find the number of miles between Janice's house and her best friend's house.

Express your answer as an integer, as a simplified fraction, or as a decimal number rounded to two places.

Writing & Thinking

13. **a.** Is the expression $6 + 3 = 9$ an equation? Explain.

  b. Is 4 a solution to the equation $5 + x = 10$? Explain.

2.2 Solving Multi-Step Linear Equations

Procedure for Solving Linear Equations that Simplify to the Form *ax + b = c*

1. Combine _____ on both sides of the equation.

2. Use the **addition principle of equality** and _____.

3. Use the **multiplication** (or **division**) **principle of equality** to _____ _____ (or _____ _____). The coefficient of the variable will become _____.

4. Check your answer by _____ in the original equation.

▶ Watch and Work

Watch the video for the example in the software and follow along in the space provided.

Example 2: Solving Linear Equations of the Form *ax + b = c*

Solve the equation: $-26 = 2y - 14 - 4y$

Solution

2.2 Solving Multi-Step Linear Equations

✏ Now You Try It!

Use the space provided to work out the solution to the next example.

Example A: Solving Linear Equations of the Form $ax+b=c$

Solve the equation: $-18 = 2y - 8 - 7y$

Solution

Remember that the objective is to get the _____ with a coefficient of $+1$.

> **Procedure for Solving Linear Equations that Simplify to the Form $ax + b = cx + d$**
>
> 1. Simplify each side of the equation by _____ and _____ on both sides of the equation.
>
> 2. Use the **addition principle of equality** and add the opposite of a _____ so that _____ are on one side and _____ are on the other side.
>
> 3. Use the **multiplication** (or **division**) **principle of equality** _____ (or _____). The coefficient of the variable will become _____.
>
> 4. Check your answer _____ in the original equation.

Every linear equation is a _____ **equation.**

Three Types of Equations	
Type of Equation	**Number of Solutions**
Conditional	_____
Identity	_____
Contradiction	_____

2.2 Exercises

Concept Check

True/False. Determine whether each statement is true or false. If a statement is false, explain how it can be changed so the statement is true. (**Note:** There may be more than one acceptable change.)

1. The first step in solving $2x + 3 = 9$ is to add 3 to both sides.

2. To solve an equation that has been simplified to $4x = 12$, you need to multiply both sides by $\frac{1}{4}$, or divide both sides by 4.

3. If an equation has no solution, it is called an identity.

4. The most general form of a linear equation is $ax + b = cx + d$.

Practice

5. Solve the following equation. Express your answer as an integer, simplified fraction, or decimal rounded to two decimal places.

   $4z - 2 = -6$

6. Solve the following equation. Express your answer as an integer, simplified fraction, or decimal rounded to two decimal places.

   $-10 = 4x - 14$

7. Solve the following equation. Express your answer as an integer, simplified fraction, or decimal rounded to two decimal places.

   $\frac{3}{8}z - \frac{8}{3}z - \frac{11}{3} = \frac{11}{12}$

8. Solve the linear equation and simplify your answer. Express your solution as an integer, a simplified fraction, or a decimal rounded to two decimal places.

$8x - 1 = 5x + 11$

9. Solve the linear equation and simplify your answer. Express your solution as an integer, a simplified fraction, or a decimal rounded to two decimal places.

$-3 + (3n + 2) = -3(-3n + 9) + 2$

10. Solve the linear equation and simplify your answer. Express your solution as an integer, a simplified fraction, or a decimal rounded to two decimal places.

$0.5u - 1.5 = 2.7u + 2.9$

11. Determine if the following equation is a conditional equation, an identity, or a contradiction.

$6(x + 2) + 5x = -3(x - 4) + 14x$

12. Determine if the following equation is a conditional equation, an identity, or a contradiction.

$3(x + 4) = 8(4 - 3x) + 27x$

Applications

Solve.

13. The tickets for a figure skating performance sold out in 4 hours. If there were 25,000 tickets sold, solve the equation $25{,}000 - 4x = 0$ to find the number of tickets sold per hour.

14. A rectangular shaped park is to have a perimeter of 900 yards. If the width must be 70 yards because of a building code, solve the equation $2l + 2(70) = 900$ to determine the length of the park.

Writing & Thinking

15. Answer each question.

   a. Simplify the expression $3(x+5) + 2(x-7)$.

   b. Solve the equation $3(x+5) + 2(x-7) = 31$.

   c. How are the methods you used to answer questions a. and b. similar? How are they different?

2.3 Working with Formulas

Formulas are general rules or _____ stated _____.

We say that the formula $d = rt$ is _____ d _____ r and t. Similarly, the formula $A = \frac{1}{2}bh$ is solved for _____ in terms of _____, and the formula $P = R - C$ (profit is equal to revenue minus cost) is solved for _____ in terms of _____.

▶ Watch and Work

Watch the video for the example in the software and follow along in the space provided.

Example 6: Solving for Different Variables

Given $V = \dfrac{k}{P}$, solve for P in terms of V and k.

Solution

✏ Now You Try It!

Use the space provided to work out the solution to the next example.

Example A: Solving for Different Variables

Given $P = \frac{I}{rt}$ solve for t in terms of I, r, and P.

Solution

2.3 Exercises

Concept Check

True/False. Determine whether each statement is true or false. If a statement is false, explain how it can be changed so the statement is true. (**Note:** There may be more than one acceptable change.)

1. When using formulas, typically it does not matter if capital or lower case letters are used: $A = a, C = c$, etc.

2. If the perimeter and length are known, $P = 2l + 2w$ can be used to find the width of a rectangle.

3. Rate of interest is stated as an annual rate in percent form.

Practice

4. Solve the following formula for the indicated variable.

   $I = Prt$; solve for t.

5. Solve the following formula for the indicated variable.

   $-3x + 8y = 9$; solve for x.

Applications

Solve.

6. If an object is shot upward with an initial velocity, v_0, in feet per second (ft/s), the velocity, v, in ft/s is given by the formula $v = v_0 - 32t$, where t is time in seconds. Find the initial velocity of an object if the velocity after 2 seconds is 15 ft/s.

7. Solve the following equation. Express your answer as an integer, simplified fraction, or decimal rounded to two decimal places.

A rectangular-shaped parking lot is to have a perimeter of 506 yards. If the width must be 100 yards because of a building code, what will the length need to be?

78 Chapter 2 Algebraic Pathways: Linear Equations and Inequalities

2.4 Applications of Linear Equations

Basic Strategy for Solving Word Problems

1. READ: _____ .

2. SET UP: Draw any type of _____ that might be helpful and
   _____ .

3. SOLVE: _____ to solve the problem.

4. CHECK: Check your work and check that _____ .

▶ Watch and Work

Watch the video for the example in the software and follow along in the space provided.

Example 3: Application: Calculating Living Expenses

Joe wants to budget $\frac{2}{5}$ of his monthly income for rent. He found an apartment he likes for $800 a month. What monthly income does he need to be able to afford this apartment?

Solution

✏ Now You Try It!

Use the space provided to work out the solution to the next example.

Example A: Application: Calculating Living Expenses

Jim plans to budget $\frac{3}{7}$ of his monthly income to send his son Taylor to private school. If the school he'd like Taylor to attend costs $1200 a month, what monthly income does Jim need to be able to afford the school?

Solution

Problems involving distance usually make use of the relationship modeled by the formula _____,
where $d =$ _____, $r =$ _____, and $t =$ _____.

The _____ (or _____) of a set of numbers can be found by _____
_____ and then dividing this sum by _____

2.4 Exercises

Concept Check

True/False. Determine whether each statement is true or false. If a statement is false, explain how it can be changed so the statement is true. (**Note:** There may be more than one acceptable change.)

1. In the distance-rate-time formula $d = r \cdot t$, the value t stands for the time spent traveling.

2. The concept of average can be used to find unknown numbers.

3. The first step to solving word problems is to draw any type of figure or diagram that might be helpful.

4. Translating English phrases into algebraic expressions can be used to solve number problems.

Practice

5. If 63 is added to a number, the result is 43 less than three times the number. Find the number.

6. 27 less than twice a number is equal to the number. What is the number?

Applications

Solve.

7. A classic car is now selling for $1000 more than two times its original price. If the selling price is now $18,000, what was the car's original price?

8. On average the number of drum sets sold in Michigan each year is 96,537, which is seven times the average number of drum sets sold each year in Vermont. How many drum sets, on average, are sold each year in Vermont?

9. The Caldwells are moving across the country. Mr. Caldwell leaves 3 hours before Mrs. Caldwell. If he averages 50 mph and she averages 90 mph, how long will it take Mrs. Caldwell to overtake Mr. Caldwell?

10. Kati's mean speed on her drive home from Detroit is 60 mph. If the total trip is 435 miles, how long should she expect the drive to take? Round your answer to two decimal places, if necessary.

11. A particular style of shoes costs the retailer $85 per pair. At what price should the retailer mark them so he can sell them at a 15% discount off the original price and still make 35% profit on his cost?

12. Akshi has five exam scores of 98, 94, 84, 95, and 95 in her biology class. What score does she need on the final exam to have a mean grade of 90? Round your answer to two decimal places, if necessary. (All exams have a maximum of 100 points.)

13. A college student realized that he was spending too much money on movies. For the remaining 5 months of the year his goal is to spend a mean of $60 a month towards movies. How much can he spend in December, taking into consideration that in the other 4 months he spent $80, $115, $50, and $45, respectively? Round your answer to two decimal places, if necessary.

Writing & Thinking

The following problem is given with an incorrect answer. Explain how you can tell that the answer is incorrect without needing to solve the problem or do any algebra; then, solve the problem correctly.

14. The perimeter of an isosceles triangle is 16 cm. Since the triangle is isosceles, two sides have the same length; the third side is 2 cm shorter than one of the two equal sides. Find the length of one of the two equal sides.
    Incorrect answer: 9 cm

2.5 Ratios, Rates, and Proportions

Ratios

A **ratio** is a _____ . The ratio of a to b can be written as

_____ or _____ or _____ .

Ratios have the following characteristics.

1. Ratios can be _____ .

2. The common units in a ratio can be _____ , just as factors can be canceled in fractions.

3. Generally, ratios are written with _____ , and the fraction is reduced to _____ .

A **rate** is a ratio with _____ . The units do not _____ .

A rate with a 1 in the denominator is called a _____ .

Changing Rates to Unit Rates

To make a rate a unit rate, _____ so the denominator is _____ .

Proportions

A **proportion** is a statement that _____ .

In symbols, _____ is a proportion.

A proportion is true if the _____ are equal.

To Solve a Proportion

1. Find the _____ (or **cross multiply**) and then
   _____ .

2. Divide both sides of the equation by the _____ .

3. _____ .

▶ Watch and Work

Watch the video for the example in the software and follow along in the space provided.

Example 8: Solving Proportions

Find the value of x if $\dfrac{4}{8} = \dfrac{5}{x}$.

Solution

✏ Now You Try It!

Use the space provided to work out the solution to the next example.

Example A: Solving Proportions

Find the value of x if $\dfrac{12}{x} = \dfrac{9}{15}$.

Solution

Solve an Application Using a Proportion

1. Identify the _____ and use a _____.

2. Set up a _____.
   (Make sure that the units are labeled so they can be seen to be in the right order.)

3. _____.

2.5 Exercises

Concept Check

True/False. Determine whether each statement is true or false. If a statement is false, explain how it can be changed so the statement is true. (**Note:** There may be more than one acceptable change.)

1. The units in the numerator and denominator of a ratio must be the same, or need to be able to be converted to the same units.

2. To make a unit rate, divide the numerator by the denominator.

3. A proportion is a statement that two ratios are being multiplied.

4. When using proportions to solve a word problem, there is only one correct way to set up the proportion.

Practice

5. Write the following rate as a simplified fraction.

   65 dogs to 26 cats

6. Write the following ratio as a fraction in lowest terms.

   10 to 15

7. Write the following ratio as a fraction in lowest terms.

   $\frac{7}{4}$ to $\frac{3}{8}$

8. Find the unit price of each of the following items. Round your answer to the nearest tenth.

   boxed rice

   16 oz at $1.75; 32 oz at $2.99

9. Find the unit price of each of the following items. Round your answer to the nearest tenth.

   noodles

   42 oz at $2.66; 14 oz at $1.05

10. Determine whether the following proportion is true or false.

    $\frac{3}{8} = \frac{12}{32}$

11. Determine whether the following proportion is true or false.

    $\frac{1}{3} = \frac{33}{100}$

12. Solve the following proportion.

$$\frac{40}{280} = \frac{x}{168}$$

13. Solve the following proportion.

$$\frac{13}{x} = \frac{36}{324}$$

Applications

Solve.

14. The distance to your brother's house is 296 miles, and the distance to Dallas is 444 miles. If it took 8 hours to drive to your brother's house, how long would you estimate the drive to Dallas to take?

15. Greg is investing his money. He thinks that he should make $13 for every $100 he invests. How much does he expect to make on an investment of $6900?

Writing & Thinking

16. In your own words, clarify how you can know that a proportion is set up correctly or not.

2.6 Modeling using Variation

Direct Variation

A variable quantity y **varies directly as** (or is **directly proportional to**) a variable x if there is a constant k such that

The constant k is called the _____.

Inverse Variation

A variable quantity y **varies inversely as** (or is **inversely proportional to**) a variable x if there is a constant k such that

The constant k is called the _____.

If a variable varies either directly or inversely with more than one other variable, the variation is said to be _____.

If the combined variation is all direct variation (the variables are multiplied), then it is called _____.

▶ Watch and Work

Watch the video for the example in the software and follow along in the space provided.

Example 6: Application: More Variation

The distance an object falls varies directly as the square of the time it falls (until it hits the ground and assuming little or no air resistance). If an object fell 64 feet in 2 seconds, how far would it have fallen by the end of 3 seconds?

Solution

2.6 Modeling Using Variation

✏️ Now You Try It!

Use the space provided to work out the solution to the next example.

Example A: More Variation

The distance an object falls varies directly as the square of the time it falls. If an object fell 64 feet in 2 seconds, how many feet will it fall in 5 seconds?

Solution

2.6 Exercises

Concept Check

True/False. Determine whether each statement is true or false. If a statement is false, explain how it can be changed so the statement is true. (**Note:** There may be more than one acceptable change.)

1. The number of hamburgers eaten varies inversely with calories consumed.

2. The equation $y = \dfrac{k}{x}$ represents direct variation.

3. Distance and time varies directly, which means they are directly proportional.

4. The circumference of a circle varies directly with its radius.

Practice

5. If y varies directly as x and $y = 74$ when $x = 13$, find y if $x = 45$. (Round your answer to the nearest hundredth.)

6. If y varies inversely as x and $y = -86$ when $x = 9$, find y if $x = 32$. (Round off your answer to the nearest hundredth.)

7. z is jointly proportional to x^3 and y^2. If $z = 106$ when $x = 7$ and $y = 4$, find z when $x = 6$ and $y = 6$. (Round off your answer to the nearest hundredth.)

Applications

Solve.

8. The distance that a free falling object falls is directly proportional to the square of the time it falls (before it hits the ground). If an object fell 93 ft in 8 seconds, how far will it have fallen by the end of 9 seconds? (Leave the variation constant in fraction form or round to at least 2 decimal places. Round your final answer to the nearest foot.)

9. The gravitational force, F, between an object and the Earth is inversely proportional to the square of the distance from the object to the center of the Earth. If an astronaut weighs 193 pounds on the surface of the Earth, what will this astronaut weigh 200 miles above the Earth? Assume that the radius of the Earth is 4000 miles. (Round off your answer to the nearest pound.)

10. The resistance, R, of a wire varies directly as its length and inversely as the square of its diameter. If the resistance of a wire 1700 ft long with a diameter of 0.13 inches is 1414 ohms, what is the resistance of 2900 ft of the same type of wire with a diameter of 0.32 inches? (Leave k in fraction form or round to at least 3 decimal places. Round off your final answer to the nearest hundredth.)

Writing & Thinking

11. Explain, in your own words, the meaning of the following terms.

    a. Direct variation

    b. Joint variation

c. Inverse variation

d. Combined variation

Discuss an example of each type of variation that you have observed in your daily life.

2.7 Solving Linear Inequalities in One Variable

The set of all real numbers between a and b is called an _____.

Types of Intervals			
Type of Interval	**Algebraic Notation**	**Interval Notation**	**Graph**
Open Interval	$a < x < b$	_____	←——(———)——→ a ⠀⠀⠀ b
Closed Interval	_____	$[a, b]$	←——[———]——→ a ⠀⠀⠀ b
Half-open Interval	$\begin{cases} a \leq x < b \\ a < x \leq b \end{cases}$	$[a, b)$ $(a, b]$	←——[———)——→ a ⠀⠀⠀ b ←——(———]——→ a ⠀⠀⠀ b
_____	$\begin{cases} x > a \\ x < b \end{cases}$	(a, ∞) $(-\infty, b)$	←——(——→ a ←——)——→ ⠀⠀⠀ b
_____	$\begin{cases} x \geq a \\ x \leq b \end{cases}$	$[a, \infty)$ $(-\infty, b]$	←——[——→ a ←——]——→ ⠀⠀⠀ b

A **solution** to an inequality is any number that _____

_____.

2.7 Solving Linear Inequalities in One Variable

The set of all solutions to an inequality is called the _____.

Addition Principle for Solving Linear Inequalities

If A and B are algebraic expressions and C is a real number, then the inequalities

$$\text{_____}$$

and

$$\text{_____}$$

are equivalent.

(If a real number is added to both sides of an inequality, _____ _____.)

Multiplication Principle for Solving Linear Inequalities

If A and B are algebraic expressions and C is a _____ real number, then the inequalities

$$\text{_____}$$

and

$$\text{_____}$$

are equivalent.

If A and B are algebraic expressions and C is a _____ real number, then the inequalities

$$\text{_____}$$

and

$$\text{_____}$$

are equivalent.

(In other words, if both sides are multiplied by a _____ number, then the sense of the inequality is _____.)

Steps for Solving Linear Inequalities

1. Combine _____ on each side of the inequality symbol.

2. Use the addition principle of inequality to _____
   _____ so that constants are on one side and variables on the other.

3. Use the multiplication (or division) principle of inequality to _____
   _____ (or _____
   _____). The coefficient of the variable will become _____ **If this coefficient is** _____ **, be sure to reverse the sense of the inequality.**

4. A quick (and generally satisfactory) check is to _____
   _____. If the statement is false, you need to look for an error in your solution.

▶ Watch and Work

Watch the video for the example in the software and follow along in the space provided.

Example 9: Solving Linear Inequalities

Solve the inequality $6x + 5 \leq -1$ and graph the solution set. Write the solution set using interval notation.

Solution

✏️ Now You Try It!

Use the space provided to work out the solution to the next example.

Example A: Solving Linear Inequalities

Solve the inequality $4x + 8 > -16$ and graph the solution set. Write the solution set using interval notation.

Solution

2.7 Exercises

Concept Check

True/False. Determine whether each statement is true or false. If a statement is false, explain how it can be changed so the statement is true. (**Note:** There may be more than one acceptable change.)

1. If only one end-point is included in an interval, it is called a half-open interval.

2. When both sides of a linear inequality are multiplied by a negative constant, the sense of the inequality should stay the same.

3. To check the solution set of a linear inequality, every solution in the solution set must be checked in the original inequality.

94   Chapter 2   Algebraic Pathways: Linear Equations and Inequalities

4. The infinity symbol ∞ does not represent a specific number.

Practice

5. Graph the interval on a real number line.

$(-\infty, 1)$

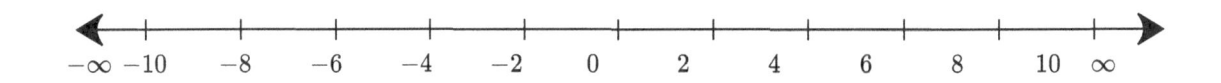

6. Consider the following inequality.

$-7 \leq x < 6$

a. Graph the solution to the given inequality.

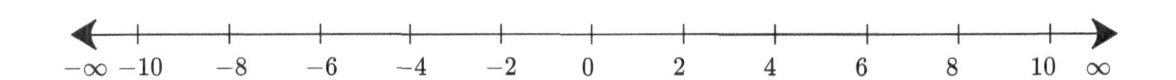

b. What type of interval does the given inequality represent?

7. Consider the following inequality.

$x + 5 \leq 3$

a. Write the solution using interval notation.

b. Graph the solution to the given inequality.

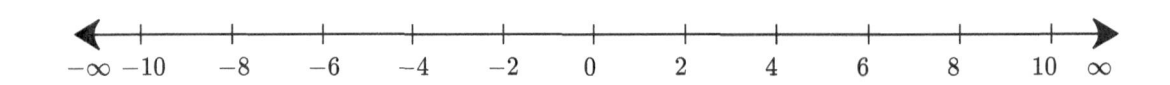

8. Consider the following inequality.

$-2x < 18$

a. Write the solution using interval notation.

b. Graph the solution to the given inequality.

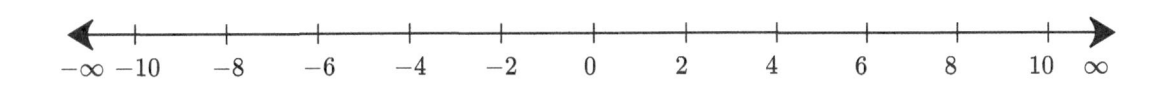

9. Consider the following inequality:

$4z - 3 \leq 2z - 7$

a. Write the solution using interval notation.

b. Graph the solution to the given inequality.

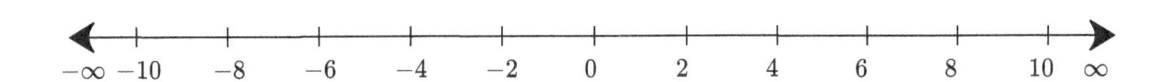

10. Consider the following inequality:

$-2(2x + 3) \geq -2x - 2$

a. Solve the linear inequality for the given variable. Simplify and express your answer in algebraic notation.

b. Graph the solution to the given inequality.

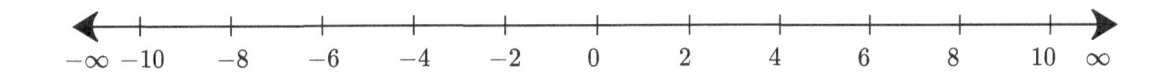

11. Consider the following inequality:

$-10 < -2z + 4 < -2$

a. Write the solution using interval notation.

b. Graph the solution to the given inequality.

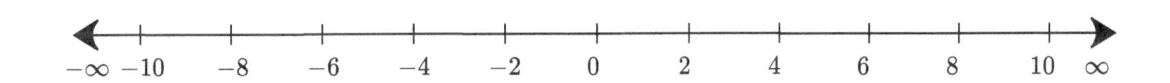

12. Consider the following inequality:

$$-3 \leq \frac{3}{4}z + 3 \leq 0$$

   a. Solve the linear inequality for the given variable. Simplify and express your answer in algebraic notation.

   b. Graph the solution to the given inequality.

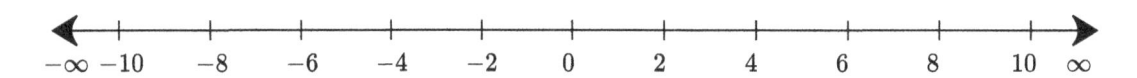

Applications

Solve.

13. To receive a C grade, a student must average more than 77 but less than or equal to 85. If Ethan received a C in the course and had five grades of 74, 80, 73, 67, and 85 before taking the final exam, what were the possible grades for his final if there were 100 points possible? (Assume all six grades are weighted equally and let x represent the unknown score.)

14. The sum of the lengths of any two sides of a triangle must be greater than the third side. If a triangle has one side that is 17 inches and a second side that is 1 inch less than twice the third side, what are the possible lengths for the second and third sides?

Writing & Thinking

15. a. Write a list of three situations where inequalities might be used in daily life.

    b. Illustrate theses situations with algebraic inequalities and appropriate numbers.

Chapter 3
Algebraic Pathways: Graphing Linear Equations and Inequalities

3.1 The Cartesian Coordinate System, Scatter Plots, and Linear Equations

3.2 Slope-Intercept Form

3.3 Point-Slope Form

3.4 Introduction to Functions and Function Notation

3.5 Linear Correlation and Regression

3.6 Systems of Linear Equations in Two Variables

3.7 Graphing Linear Inequalities in Two Variables

Name: _____ Date: _____ 99

3.1 The Cartesian Coordinate System, Scatter Plots, and Linear Equations

We say that $(2, 7)$ is **a** _____ the equation. The first number, 2, is called the _____ (or _____), and the second number, 7, is called the _____ (or _____).

The **Cartesian coordinate system** relates algebraic equations and ordered pairs of real numbers to geometry in a plane. In this system, two number lines intersect at a right angle and separate the plane into four _____. The **origin**, designated by the ordered pair $(0, 0)$, is _____. The horizontal number line is called the _____ or _____. The vertical number line is called the _____ or _____.

One-to-One Correspondence

There is a **one-to-one correspondence** between _____.

In statistics, data is sometimes given in the form of ordered pairs, called _____, where each ordered pair represents two pieces of information about one situation. The ordered pairs are plotted on a graph and the graph is called a _____, or _____.

Solution Set of an Equation in Two Variables

The **solution set** of an equation in two variables, x and y, consists of _____.

Standard Form of a Linear Equation

Any equation of the form

_____,

where A, B, and C are real numbers and A and B are not both equal to 0, is called the **standard form of a linear equation**.

To Graph a Linear Equation in Two Variables

1. Locate any two points that _____.

2. _____.

3. _____.

4. **To check:** Locate a third point that _____.

▶ Watch and Work

Watch the video for the example in the software and follow along in the space provided.

Example 6: Graphing a Linear Equation in Two Variables

Graph: $x - 2y = 1$

Solution

✏️ Now You Try It!

Use the space provided to work out the solution to the next example.

Example A: Graphing a Linear Equation in Two Variables

Graph: $2x - y = 2$.

Solution

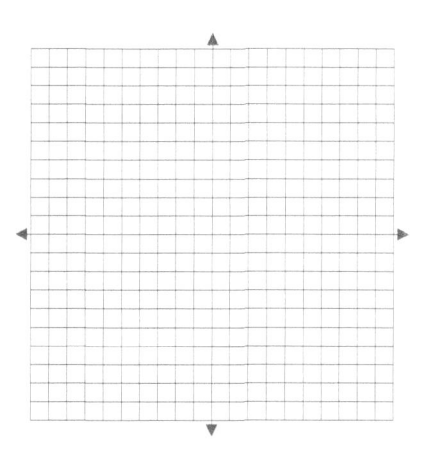

Intercepts

1. To find the **y-intercept** (where the line crosses the y-axis), _____ .

2. To find the **x-intercept** (where the line crosses the x-axis), _____ .

3.1 Exercises

Concept Check

True/False. Determine whether each statement is true or false. If a statement is false, explain how it can be changed so the statement is true. (**Note:** There may be more than one acceptable change.)

1. The y-intercept is the point where a line crosses the y-axis.

2. The terms ordered pair and point are used interchangeably.

3. A horizontal line does not have a *y*-intercept.

4. All *x*-intercepts correspond to an ordered pair of the form $(0, y)$.

Practice

5. Consider the following.

A $(2, -1)$

a. Plot the given point on the graph.

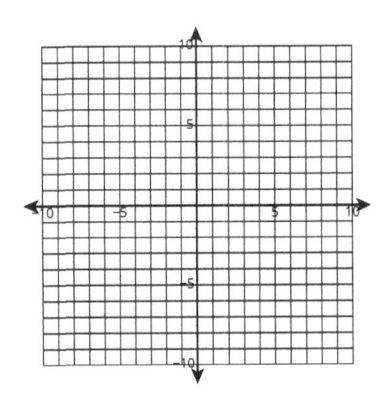

b. Identify the coordinates of the point A on the graph.

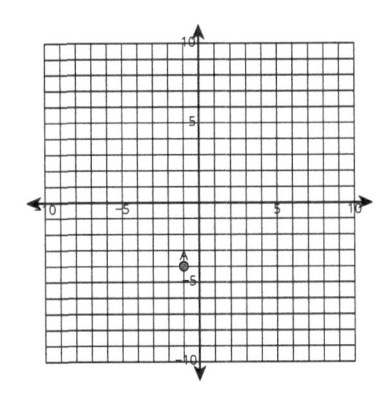

6. Consider the following.

A $(-5, -4)$ B $(-1, 1)$

   a. Plot the given points on the graph.

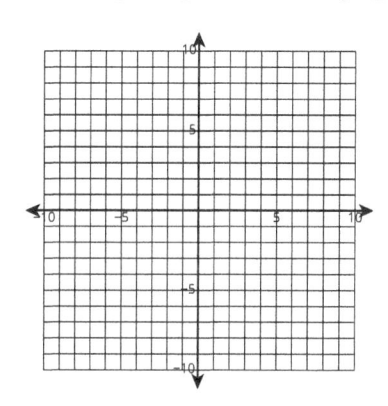

   b. Identify the coordinates of the points A and B on the graph.

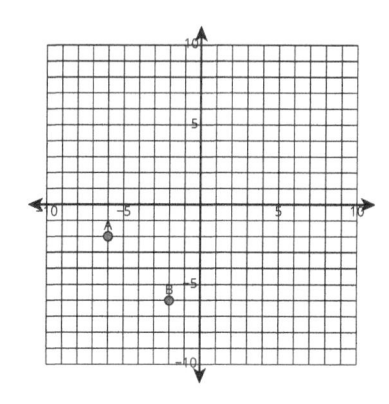

7. Consider the following equation.

$5x + 5y = 18$

   a. Determine the missing coordinate in the ordered pair $(2, ?)$ so that it will satisfy the given equation.

   b. Determine the missing coordinate in the ordered pair $(?, 3)$ so that it will satisfy the given equation.

8. Graph the line by plotting any two ordered pairs with integer value coordinates that satisfy the equation.

$-9x + 9y = 0$

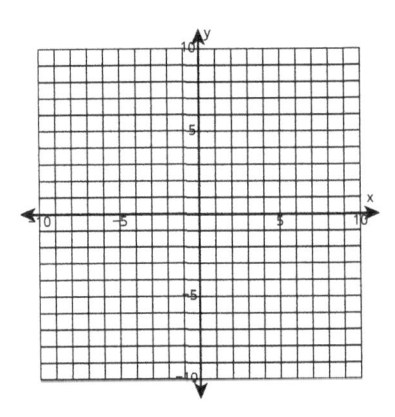

9. Graph the line by plotting any two ordered pairs that satisfy the equation.

$y = -\dfrac{1}{5}x - 1$

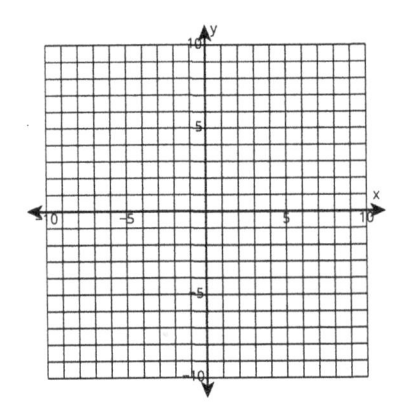

10. Find the *y*-intercept and *x*-intercept of the following linear equation.

$-10x + 5y = 50$

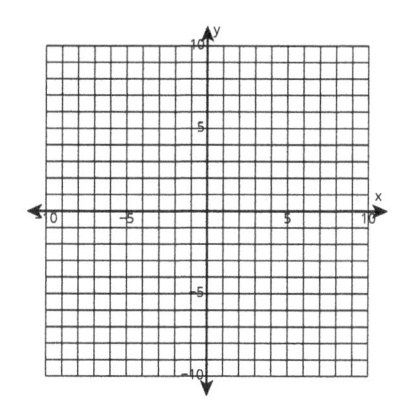

11. Find the *y*-intercept and *x*-intercept of the following linear equation.

$-7x + 6y = 16$

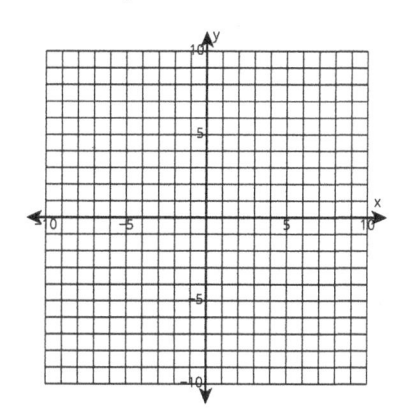

106 Chapter 3 Algebraic Pathways: Graphing Linear Equations and Inequalities

Applications

Solve.

12. The following table gives the average number of hours 7 junior high students were left unsupervised each day and their corresponding overall grade averages.

Hours Unsupervised	0	0.5	1	2.5	3.5	4.5	5
Overall Grades	96	92	88	80	76	72	68

Draw a scatter plot of the given data.

13. Consider the equation $V = 4h$ where V is the volume (in cubic centimeters) of a box with a variable height h in centimeters and a fixed base of area 4 cm^2.

  a. Complete the table below so that each ordered pair will satisfy the equation.

  b. Plot the points corresponding to the ordered pairs from the previous part.

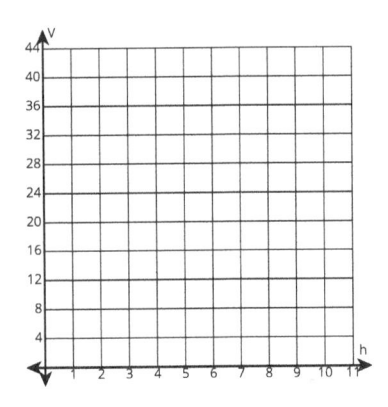

14. To mail a letter to Vienna, Austria the post office charges a flat rate of $6.50 and an additional $0.25 for every ounce the letter weighs. The cost of mailing a letter is determined by the equation $y = 0.25x + 6.50$, where y is the cost of the postage and x is the weight of the letter in ounces.

   a. Graph the equation by finding two points that satisfy the equation and plotting them on the graph. To plot a point, click on the appropriate position on the graph. You may move a point by clicking and dragging.

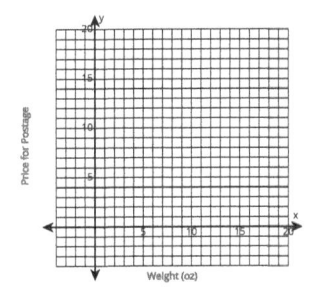

   b. Use the graph to estimate to the nearest quarter of a dollar ($0.25) the cost to send a letter that weighs 4 ounces.

3.2 Slope-Intercept Form

For a line, the _____ is called the **slope of the line**.

> **Slope**
> Let $P_1(x_1, y_1)$ and $P_2(x_2, y_2)$ be two points on a line. The **slope** can be calculated as follows.
>
> _____
>
> **Note:** _____ is standard notation for representing the slope of a line.

▶ Watch and Work

Watch the video for the example in the software and follow along in the space provided.

Example 2: Finding the Slope of a Line

Find the slope of the line that contains the points $(1, 3)$ and $(5, 1)$, and then graph the line.

Solution

3.2 Slope-Intercept Form   109

✏️ Now You Try It!

Use the space provided to work out the solution to the next example.

Example A: Finding the Slope of a Line

Find the slope of the line that contains the points (0,5) and (4,2), and then graph the line.

Solution

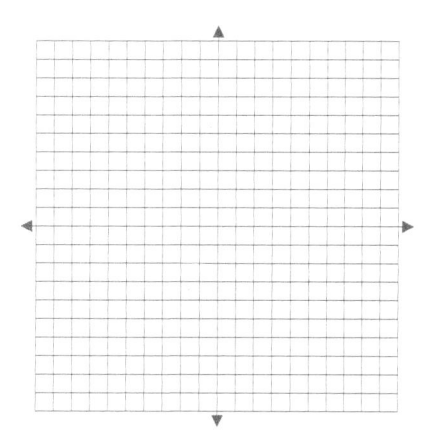

Positive and Negative Slopes

Lines with **positive slope** go
_____.

Lines with **negative slope** go
_____.

Horizontal and Vertical Lines

The following two general statements are true for horizontal and vertical lines.

1. For **horizontal lines** (of the form _____.

2. For **vertical lines** (of the form _____.

Slope-Intercept Form

_____ is called the **slope-intercept form** for the equation of a line, where m is the **slope** and $(0, b)$ is the **y-intercept**.

3.2 Exercises

Concept Check

True/False. Determine whether each statement is true or false. If a statement is false, explain how it can be changed so the statement is true. (**Note:** There may be more than one acceptable change.)

1. If the *y*-intercept and the slope of a line are given, there is enough information to write the equation of the line.

2. When using the slope formula, the slope of a line changes if the order of the points is reversed.

3. A line that falls (decreases) from left to right has a negative slope.

4. The line that represents the equation $y = 2x + 4$ has a *y*-intercept of $(0, 4)$.

Practice

5. Find the slope determined by the following pair of points.

   $(8, 1), (-5, 4)$

6. Find the slope determined by the following pair of points.

   $\left(\frac{11}{4}, -2\right), \left(3, \frac{5}{2}\right)$

7. Consider the following linear equation.

   $y = -8$

   a. Identify if the given line is horizontal or vertical and give its slope.

b. Graph the line by plotting two points.

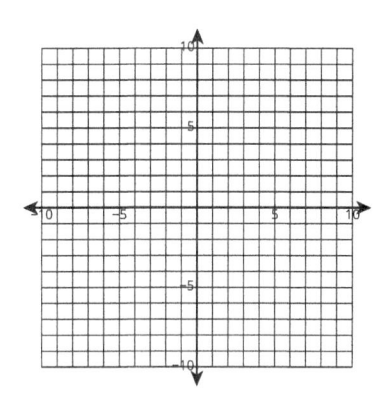

8. Consider the following linear equation.

$x = 9$

   a. Identify if the given line is horizontal or vertical and give its slope.

   b. Graph the line by plotting two points.

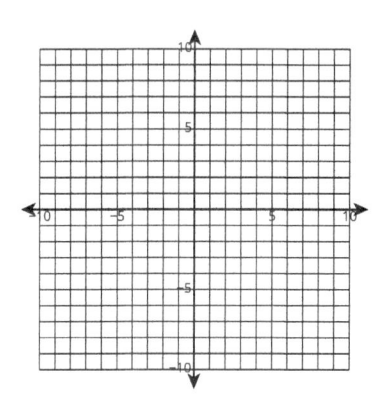

9. Consider the following linear equation.

$y = 4 - \dfrac{1}{2}x$

   a. Determine the slope and the *y*-intercept (entered as an ordered pair) of the equation above. Reduce all fractions to lowest terms.

b. Graph the line.

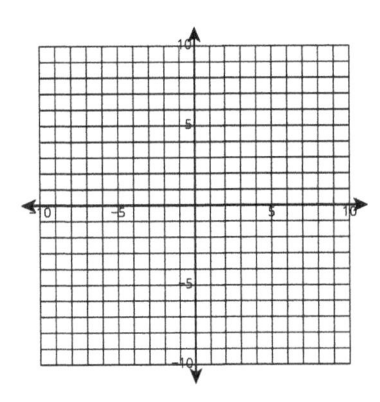

10. Consider the following linear equation.

    $2x + 8y = 32$

    a. Determine the slope and the *y*-intercept (entered as an ordered pair) of the equation above. Reduce all fractions to lowest terms.

    b. Graph the line.

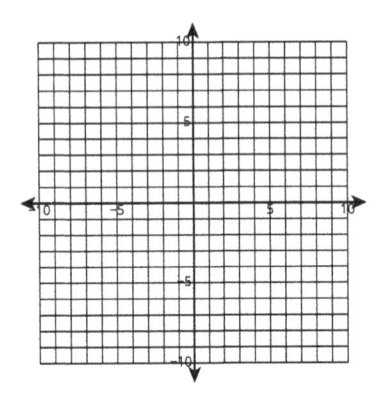

11. Consider the following *y*-intercept and slope:

    $(0, 1), m = 2$

    a. Find the equation of the line for the *y*-intercept and slope given above.

b. Determine the graph of the line with the given *y*-intercept and slope.

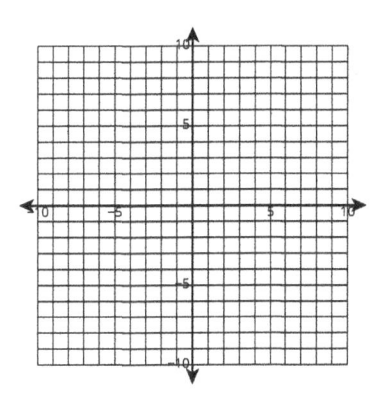

Applications

Solve.

12. The percent of U.S. citizens that use the Internet on a daily basis is shown on the graph, where the year 2000 corresponds to $x = 0$. Use the concepts of *y*-intercept and slope of the line to find the equation that best models the given data.

Enter your answer in slope-intercept form.

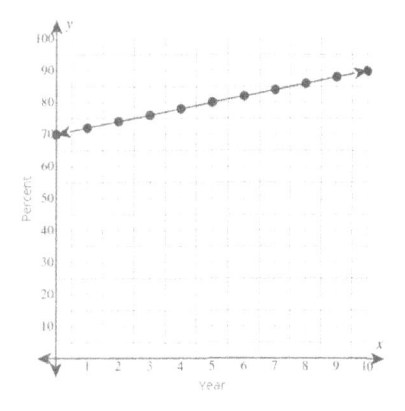

13. In the year 2008, Chris's boat had a value of $22,000. When he bought the boat in 2004 he paid $27,000. If the value of the boat depreciated linearly, what was the annual rate of change of the boat's value? Round your answer to the nearest hundredth if necessary.

Writing & Thinking

14. **a.** Explain in your own words why the slope of a horizontal line must be 0.

b. Explain in your own words why the slope of a vertical line must be undefined.

3.3 Point-Slope Form

Point-Slope Form

An equation of the form

_____

is called the **point-slope form** for the equation of a line that

_____ .

Finding the Equation of a Line Given Two Points

To find the equation of a line given two points on the line:

1. Use the formula
_____.

2. Use this slope, m, and _____

_____.

▶ Watch and Work

Watch the video for the example in the software and follow along in the space provided.

Example 4: Finding Equations of Lines Using a Graph

Write an equation in standard form for the following line.

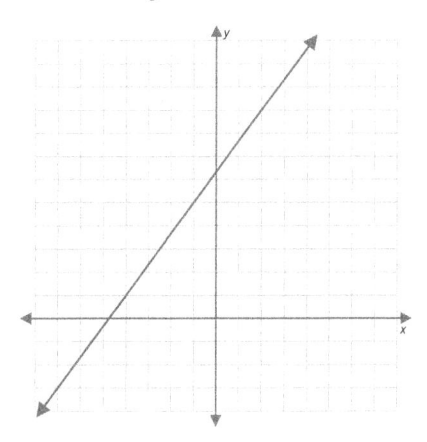

Solution

✏️ Now You Try It!

Use the space provided to work out the solution to the next example.

Example A: Finding Equations of Lines Using a Graph

Write an equation in standard form for the following line.

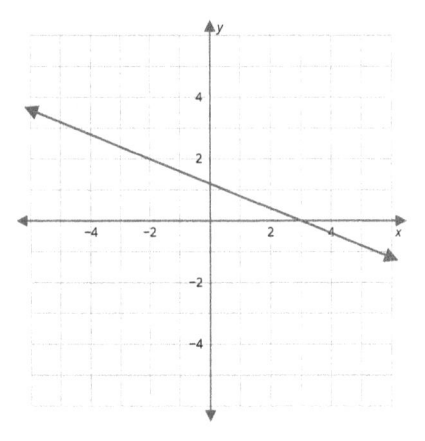

Solution

Parallel and Perpendicular Lines

Parallel lines are lines that _____ and these lines have the _____.

Perpendicular lines are lines that _____ and whose slopes are _____. Horizontal lines are perpendicular to _____.

3.3 Exercises

Concept Check

True/False. Determine whether each statement is true or false. If a statement is false, explain how it can be changed so the statement is true. (**Note:** There may be more than one acceptable change.)

1. Given two perpendicular lines (neither of which have slope 0), we know that one has a positive slope and the other has a negative slope.

2. If Line 2 is parallel to Line 3, then the slope of Line 2 equals the slope of Line 3.

3. A line perpendicular to a horizontal line has a slope that is undefined.

4. All pairs of lines are either parallel or perpendicular.

Practice

5. Find the equation (in slope-intercept form) of the line with the given slope that passes through the point with the given coordinates.

slope: $\frac{-10}{9}$, ordered pair: $\left(0, \frac{-7}{2}\right)$

6. Find the equation (in slope-intercept form) of the line with the given slope that passes through the point with the given coordinates.

slope: 0, ordered pair: $(1, -9)$

7. Find the equation (in slope-intercept form) of the line passing through the points with the given coordinates.

$(-2, -1)$, $(0, 5)$

8. Find the equation (in slope-intercept form) of the line passing through the points with the given coordinates.

$(2, -3)$, $(4, 4)$

9. Find the equation of the line shown. Enter your answer in point-slope form.

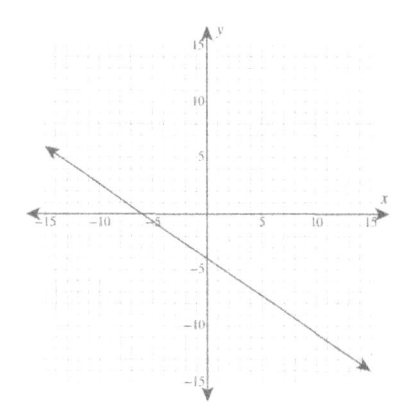

3.3　Point-Slope Form　**119**

10. Find the equation for the line that passes through the point $(2, 2)$, and that is **perpendicular** to the line with the equation $x = -1$.

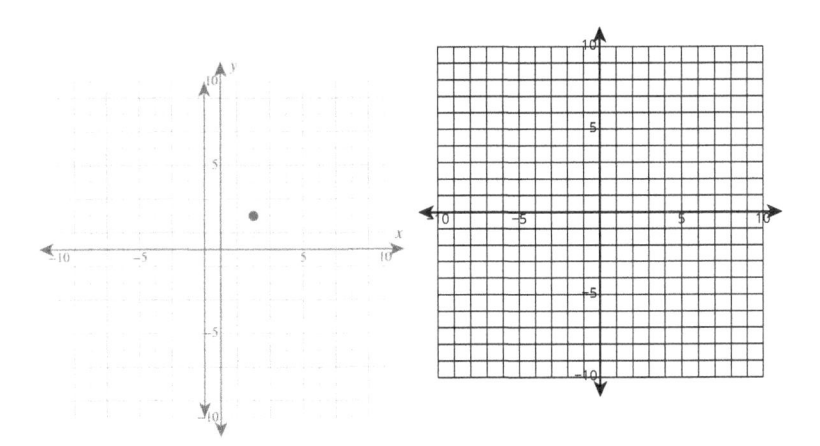

11. Find the equation for the line that passes through the point $(-4, 3)$, and that is **parallel** to the line with the equation $y = -3$.

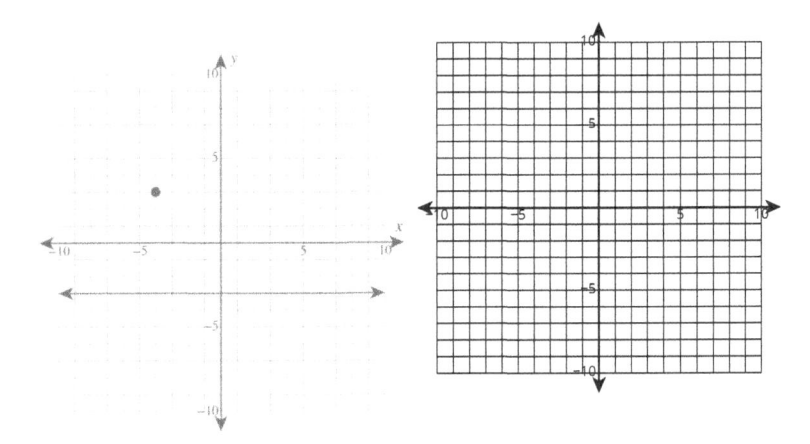

Applications

Solve.

12. The table lists the number of members at a local gym for selected years, where year 1 represents 2002, year 3 represents 2004, and so on.

Year	x	y = number of members
2002	1	43
2004	3	54
2006	5	70
2008	7	89
2010	9	97

Use the ordered pairs $(5, 70)$ and $(7, 89)$ to find the equation of a line that approximates the data. Express your answer in slope-intercept form. (If necessary, round the slope to the nearest hundredth and the y-intercept to the nearest whole number.)

13. A NYC Taxi charges a fare of $5 plus $0.25 per eighth of a mile for a ride.

   a. Find an equation for the fare f in terms of the number of miles m.

   b. Use the equation found in Part **a.** to determine the cost for a 15-mile ride to JFK airport?

Writing & Thinking

14. Ramps for persons in wheelchairs or otherwise handicapped are now built into most buildings and walkways. (If ramps are not present in a building, then there must be elevators.) What do you think that the slope of a ramp should be for handicapped access? Look in your library or contact your local building permit office to find the recommended slope for such ramps.

3.4 Introduction to Functions and Function Notation

Relation, Domain, and Range

A **relation** is a set of _____.

The **domain**, D, of a relation is the set of _____.

The **range**, R, of a relation is the set of _____.

Functions

A **function** is a relation in which _____

_____.

Vertical Line Test

If **any** vertical line intersects the graph of a relation at more than one point, then

_____.

Linear Function

A **linear function** is a function represented by an equation of the form

_____.

The domain of a linear function is _____:

_____.

▶ Watch and Work

Watch the video for the example in the software and follow along in the space provided.

Example 6: Evaluating Functions

For the function $g(x) = 4x + 5$, find:

  a. $g(2)$

  b. $g(-1)$

  c. $g(0)$

Solution

✏️ Now You Try It!

Use the space provided to work out the solution to the next example.

Example A: Evaluating Functions

For the function $g(x) = 3x - 2$ find:

a. $g(3)$

b. $g(-2)$

c. $g(0)$

Solution

3.4 Exercises

Concept Check

True/False. Determine whether each statement is true or false. If a statement is false, explain how it can be changed so the statement is true. (**Note:** There may be more than one acceptable change.)

1. If the domain of a linear function is not explicitly stated, the implied domain is the set of all values of x that produce real values for y.

2. A relation is a function in which each domain element has exactly one corresponding range element.

3. In a function, the range elements can have more than one corresponding domain element.

4. If $s = \{(1, -6), (3, 5), (4, 0), (1, 2)\}$ then s is a function.

Practice

5. Consider the following relation.

$\{(0, -8), (-4, 1), (7, 2), (6, 6), (7, -2), (-7, -2)\}$

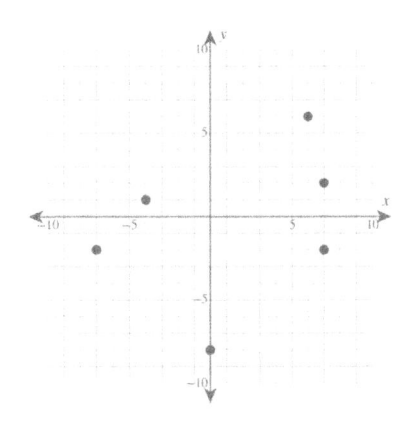

   a. Determine the values in the domain and range of the relation. Enter repeated values only once.

   b. Does the relation represent a function?

6. Consider the following graph.

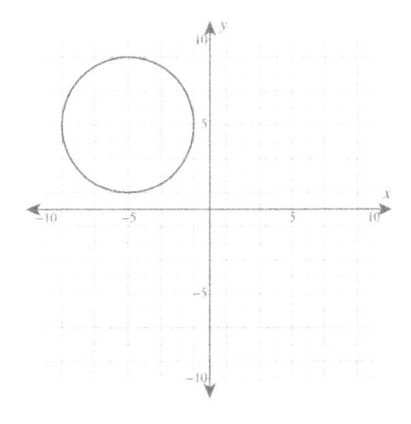

   a. Does the graph represent a function?

b. Determine the domain and range of the graph.

7. Consider the following graph.

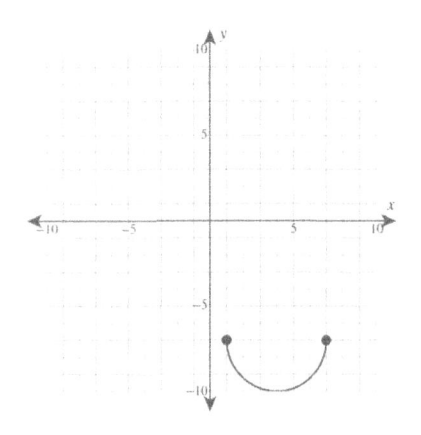

a. Does the graph represent a function?

b. Determine the domain and range of the graph.

8. Consider the following graph.

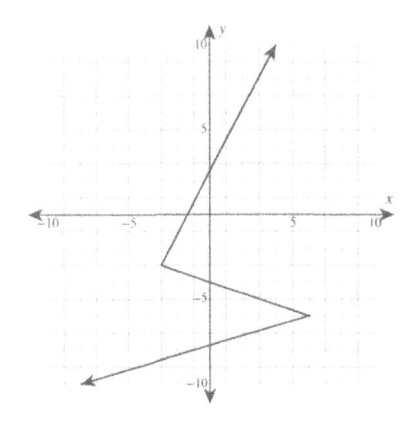

a. Does the graph represent a function?

b. Determine the domain and range of the graph.

9. State the domain of the following function, using interval notation.

$$y = \frac{5x^2 - 7x + 6}{x - 2}$$

10. State the domain of the function, using interval notation.

$8x + 8y = 25$

11. Consider the function below.

$f(x) = 4x + 9$

Find the value of $f(4.3)$.

12. Consider the function $h(x) = x^3 - 5x$. Find the value of $h(-3)$.

13. Using the graph of f, determine $f(-2)$.

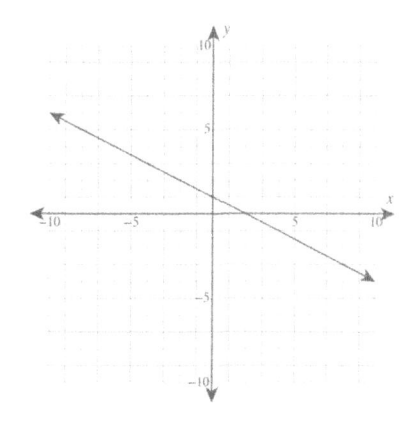

Applications

Solve.

14. A nurse hangs a 1000-milliliter IV bag which is set to drip at 120 milliliters per hour. Create a model of this situation to represent the amount of IV solution left in the bag after x hours.

   a. The y-intercept is the amount of IV solution in the bag initially (time $= 0$). What is the y-intercept?

   b. The slope is equal to the rate that the IV solution is dispensed per hour. What is the slope? (Hint: Consider whether the amount of IV solution in the bag is increasing or decreasing and how this would affect the slope.)

   c. Write an equation in slope-intercept form to model this situation.

   d. Write the equation from Part **c.** using function notation.

   e. State the domain and range of the function.

   f. State any additional restrictions that should be made on the domain for it to make sense in the context of this problem.

   g. How much IV solution is left in the bag after 5 hours?

3.5 Linear Correlation and Regression

Correlation

When a data set is presented in a scatter plot and there appears to be _____ to the data, we say that the variables represented by the data are _____. An upward trend in the data indicates a _____. A downward trend in the data indicates a _____.

▶ Watch and Work

Watch the video for the example in the software and follow along in the space provided.

Example 1: Identifying Correlations

Determine if the data set appears to have a positive correlation, a negative correlation, or no correlation.

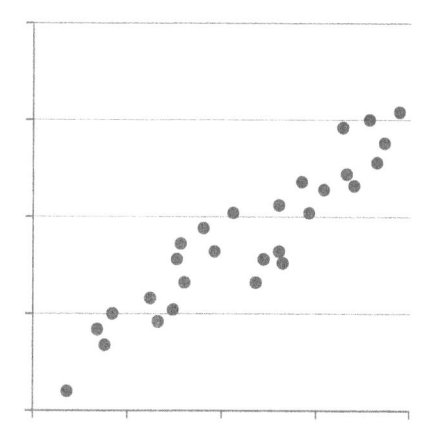

Solution

Chapter 3 Algebraic Pathways: Graphing Linear Equations and Inequalities

✏ Now You Try It!

Use the space provided to work out the solution to the next example.

Example A: Identifying Correlations

Determine if the data set appears to have a positive correlation, a negative correlation, or no correlation.

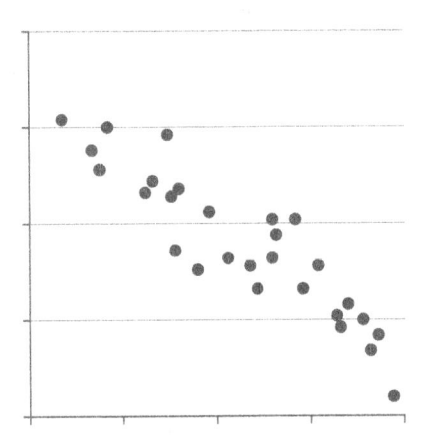

Solution

Correlation Coefficient

The **correlation coefficient**, denoted by r, is a value between _____, inclusive. A positive r indicates a positive correlation between two variables while a negative r indicates a _____. When $r = 0$, there is _____.

Perfectly Positive and Perfectly Negative Correlation

If $r = 1$, then the data falls in _____. This is called **perfectly positive correlation**.

If $r = -1$, then the data falls in a _____. This is called **perfectly negative correlation**.

Linear Regression

Linear regression is the process of _____.

Regression Line

The **regression line**, or _____ , is the line that fits the data as closely as possible. This line is represented by

_____ ,

where a is the slope of the line and b is the y-intercept.

3.5 Exercises

Concept Check

True/False. Determine whether each statement is true or false. If a statement is false, explain how it can be changed so the statement is true. (**Note:** There may be more than one acceptable change.)

1. When a data set is presented in a scatter plot and there appears to be an upward trend in the data, we say there is a positive correlation.

2. If the correlation coefficient r of a data set is close to zero, the two variables have a strong correlation.

3. If the correlation coefficient r of a data set is zero, the variables are perfectly correlated.

4. Linear regression can be used to predict some values under certain conditions.

Practice

5. Choose the graph that appears to have a positive linear correlation.

a.

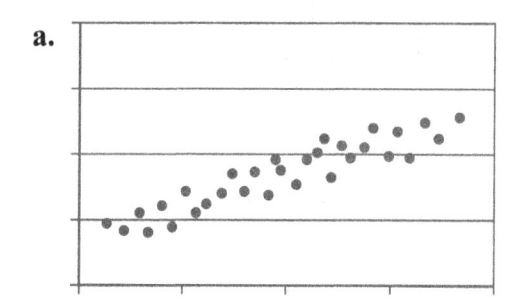

b.

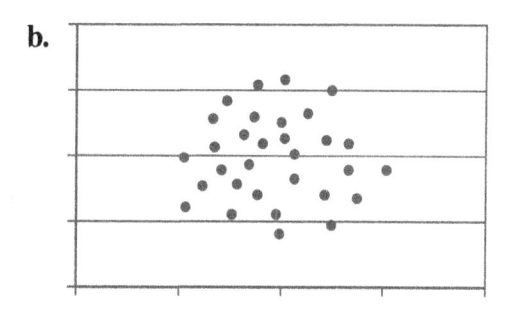

130   Chapter 3   Algebraic Pathways: Graphing Linear Equations and Inequalities

c.

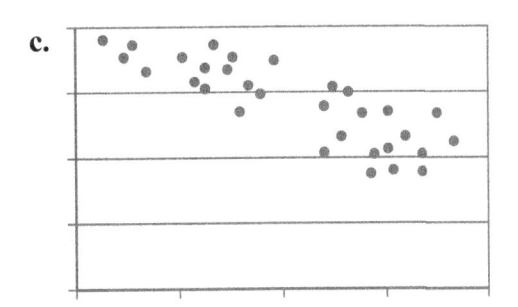

d.

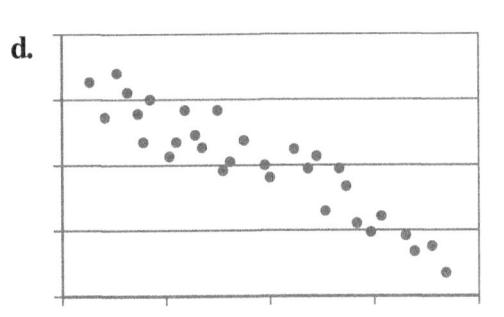

6. Choose the graph that appears to have a negative linear correlation.

a.

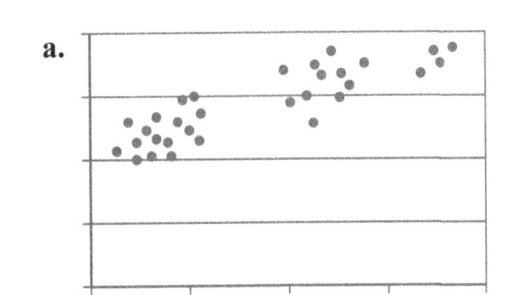

b.

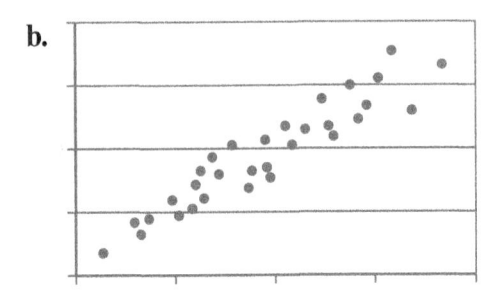

c.

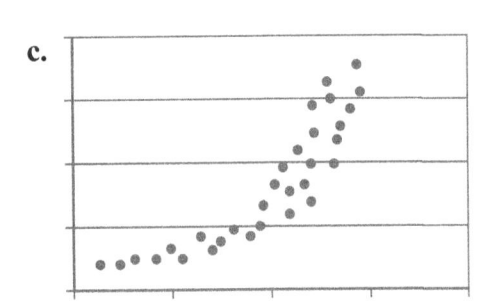

d.

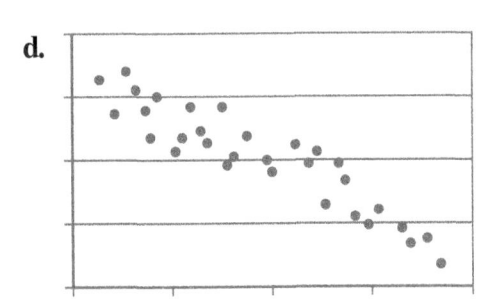

7. Use the points $(10, 40)$ and $(16, 10)$ from the following data set to determine the point-slope form of the equation that represents the data set.

x	8	10	10	11	12	14	15	16	16
y	41	40	30	32	25	22	22	14	10

8. Use the linear regression model $\hat{y} = -13.7x + 579.67$ to predict the y-value for $x = 10$.

Applications

Solve.

9. The following table gives the data for the number of outs recorded and the number of runs scored by a local t-ball team last season. Use the data from the second game and the last game to estimate the equation of the regression line, $\hat{y} = a + bx$.

T-Ball Season Stats									
Number of Outs	3	4	4	5	6	6	7	8	8	9
Number of Runs	7	7	8	10	10	12	13	15	16	17

10. The following table gives the data for the grades on the midterm exam and the grades on the final exam. Use a graphing calculator to determine the equation of the regression line, $\hat{y} = a + bx$. Round the slope and y-intercept to the nearest thousandth.

Grades on Midterm and Final Exams									
Grades on Midterm	60	73	73	96	73	87	60	91	65	80
Grades on Final	85	63	70	96	76	74	61	87	80	65

Writing & Thinking

11. Explain how the line between one pair of points in a data set can a good fit for the data while the line between a different pair of points might be a bad fit for the data.

3.6 Systems of Linear Equations in Two Variables

_____ are said to form a **system of equations** or a **set of simultaneous equations**.

Consistent and Inconsistent Systems of Linear Equations

1. A system is **consistent** if it has _____.

2. A system is **inconsistent** if it has _____.

Dependent and Independent Systems of Linear Equations

If the graphs of two linear equations are

a. the same line, then _____.

b. different lines, then _____.

To Solve a System of Linear Equations by Graphing

1. Graph both linear equations on _____.

2. Observe the point of _____.

   a. If the slopes of the two lines are different, then _____. The system has _____.

   b. If the lines have the same slope and different y-intercepts, then _____. The system has _____.

   c. If the lines are the same line, then _____. There are _____ _____ solutions.

3. Check the solution (if there is one) in both of the original equations.

To Solve a System of Linear Equations by Substitution

1. Solve one of the equations for _____ .

2. Substitute the resulting _____ .

3. Solve this new equation, if possible, and then _____
   _____ .

4. Check the solution in _____ .

To Solve a System of Linear Equations by Addition

1. Write the equations in
   _____ .

2. Multiply all terms of one equation by _____
   _____ .

3. Add the two equations by
   _____ .

4. **Back substitute into one of the original equations** to
   _____ .

5. Check the solution in _____ .

▶ Watch and Work

Watch the video for the example in the software and follow along in the space provided.

Example 6: A System with an Infinite Number of Solutions

Solve the following system by using the method of addition: $\begin{cases} 3x - \frac{1}{2}y = 6 \\ 6x - y = 12 \end{cases}$.

Solution

Now You Try It!

Use the space provided to work out the solution to the next example.

Example A: A System with an Infinite Number of Solutions

Solve the following system by using the method of addition: $\begin{cases} 6x + 3y = 15 \\ 2x + y = 5 \end{cases}$

Solution

3.6 Exercises

Concept Check

True/False. Determine whether each statement is true or false. If a statement is false, explain how it can be changed so the statement is true. (**Note:** There may be more than one acceptable change.)

1. A system of equations with graphs that are parallel lines has exactly one solution.

2. A system of equations with graphs that intersect at one point has exactly one solution.

3. The method of substitution reduces the problem from one of solving two equations in two variables to solving one equation in one variable.

4. When using the method of addition, the solution only needs to be checked in one of the original equations.

Practice

5. Solve the following system of equations by graphing. Then determine whether the system is consistent or inconsistent and whether the equations are dependent or independent. If the system is consistent, give the solution.

$$\begin{cases} x + y = 11 \\ -5x + 5y = 5 \end{cases}$$

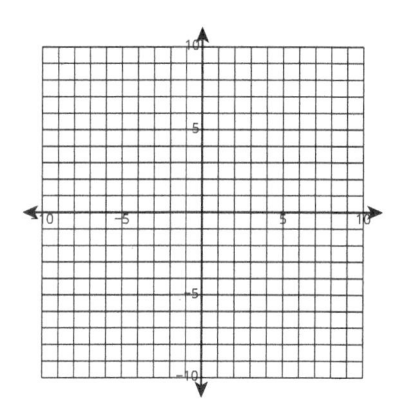

6. Solve the following system of equations by graphing. Then determine whether the system is consistent or inconsistent and whether the equations are dependent or independent. If the system is consistent, give the solution.

$$\begin{cases} 4x + y = -10 \\ 8x + 2y = -20 \end{cases}$$

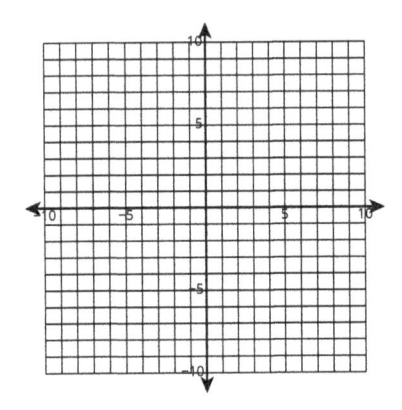

7. Solve the following system of linear equations by substitution and determine whether the system has one solution, no solution, or an infinite number of solutions. If the system has one solution, find the solution.

$$\begin{cases} -5x + y = 2 \\ 10x + 5y = 10 \end{cases}$$

8. Solve the following system of linear equations by addition. Indicate whether the given system of linear equations has one solution, has no solution, or has an infinite number of solutions. If the system has one solution, find the solution.

$$\begin{cases} 2.4x - 1.2y = -1.2 \\ 7.2x - 3.6y = -0.6 \end{cases}$$

Applications

Solve.

9. A grocer wants to mix two kinds of coffee. One kind sells for $0.90 per pound, and the other sells for $2.25 per pound. He wants to mix a total of 22 pounds and sell it for $1.60 per pound. How many pounds of each kind should he use in the new mix? (Round off the answers to the nearest hundredth.)

10. A total of $6000 is invested: part at 7% and the remainder at 15%. How much is invested at each rate if the annual interest is $710?

Writing & Thinking

11. Explain the advantages of solving a system of linear equations

   a. by graphing,

   b. by substitution.

3.7 Graphing Linear Inequalities in Two Variables

Half-plane A straight line separates _____ .

The points on one side of the line are in _____

_____ .

Boundary line _____ is called the **boundary line.**

Closed half-plane If the boundary line is _____ , then the half-plane is said to be _____ .

Open half-plane If the boundary line is _____ , then the half-plane is said to be _____ .

3.7 Graphing Linear Inequalities in Two Variables   139

Graphing Linear Inequalities

1. First, graph _____ (_____ if the inequality is _____ , _____ if the inequality is _____).

2. Next, determine which _____ .

   Method 1

   a. Test any one _____ .

   b. If the test-point satisfies the inequality, shade the _____ .

   Note: The point $(0, 0)$, if it is not on the boundary line, is usually the easiest point to test.

   Method 2

   a. Solve the inequality for y _____ .

   b. If the solution shows _____ , then shade the half-plane _____ _____ .

   c. If the solution shows _____ , then shade the half-plane _____ _____ .

   Note: If the boundary line is vertical, _____ . If the solution shows _____ , then shade the half-plane _____ . If the solution shows _____ , then shade the half-plane _____ .

3. The shaded half-plane (and the line if it is solid) is _____ .

▶ Watch and Work

Watch the video for the example in the software and follow along in the space provided.

Example 2: Graphing Linear Inequalities

Graph the solution set to the inequality $y > 2x$.

Solution

Now You Try It!

Use the space provided to work out the solution to the next example.

Example A: Graphing Linear Inequalities

Graph the solution set to the inequality $y < 3x$.

Solution

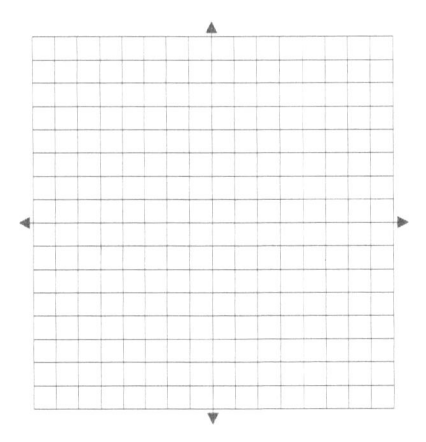

To Solve a System of Two Linear Inequalities

1. For each inequality, graph the boundary line and _____.

2. Determine the region of the graph that is _____.

   (This region is called the _____ and is the _____ of the system.)

3. To check, pick one test-point in the _____ and verify that it _____.

Note: If there is _____, then the system has no solution.

Possible Solutions to Systems of Linear Inequalities

When the boundary lines are parallel there are three possibilities:

1. The common region will be in the form of _____.

2. The common region will be _____.

3. There will be _____.

3.7 Exercises

Concept Check

True/False. Determine whether each statement is true or false. If a statement is false, explain how it can be changed so the statement is true. (**Note:** There may be more than one acceptable change.)

1. A solid boundary line indicates that the points on that line are included in the solution.

2. If the solution set is an open half-plane, then the boundary line is included in the solution.

3. When boundary lines are parallel, the system of linear inequalities has no solution.

4. Half-planes are the graphs of linear inequalities.

Practice

5. Graph the solution set of the following linear inequality:

$4x + 4y < -24$

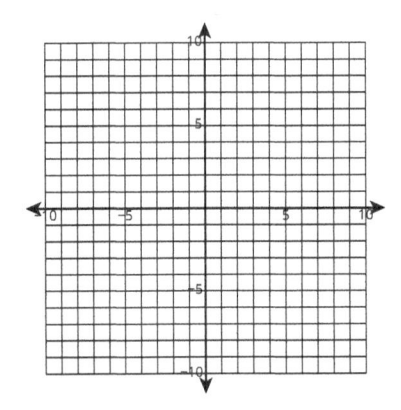

6. Graph the solution set of the following linear inequality:

$-y - 2 \leq 2x$

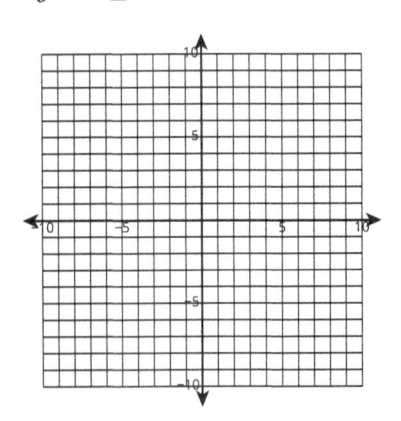

7. Graph the solution set of the following linear inequality:

$x - 7 < 0$

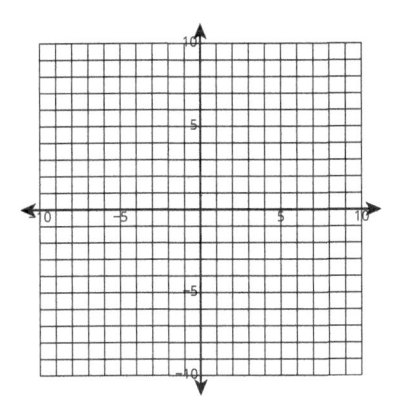

8. Solve the system of two linear inequalities graphically.

$$\begin{cases} 2x + 7y < 14 \\ x \geq -3 \end{cases}$$

   a. Graph the solution set of the **first** linear inequality.

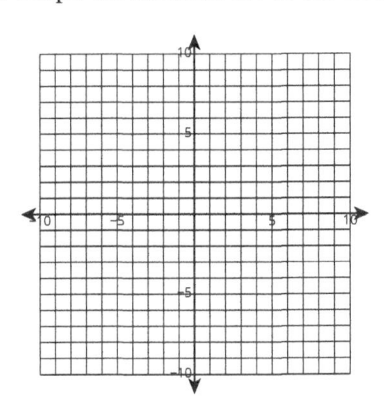

   b. Graph the solution set of the **second** linear inequality.

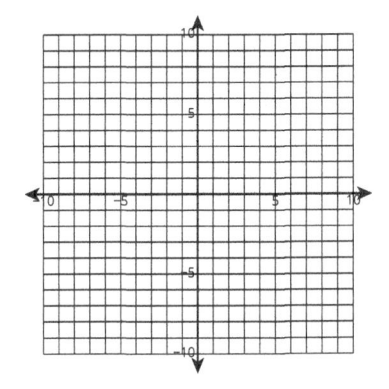

c. Find the region with points that satisfy both inequalities.

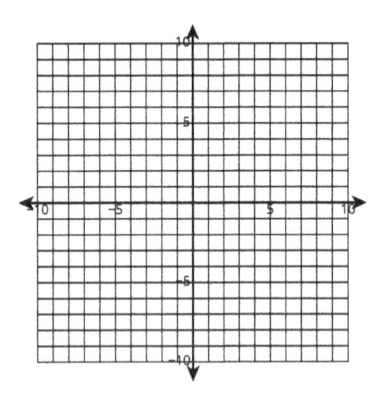

9. Solve the system of two linear inequalities graphically.

$$\begin{cases} y \leq x + 2 \\ y > -5x - 10 \end{cases}$$

a. Graph the solution set of the **first** linear inequality.

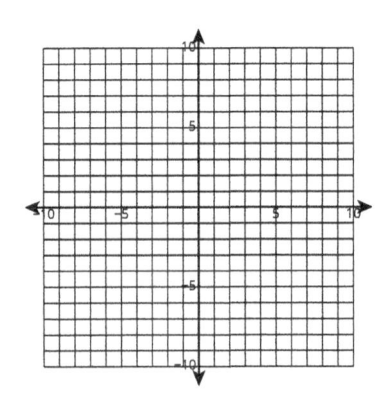

b. Graph the solution set of the **second** linear inequality.

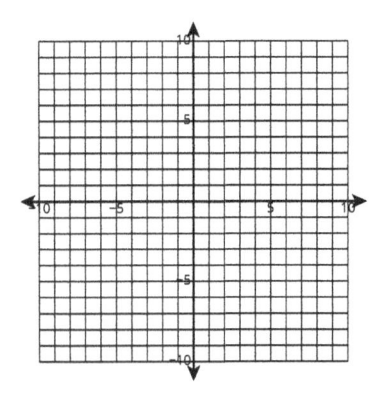

c. Find the region with points that satisfy both inequalities.

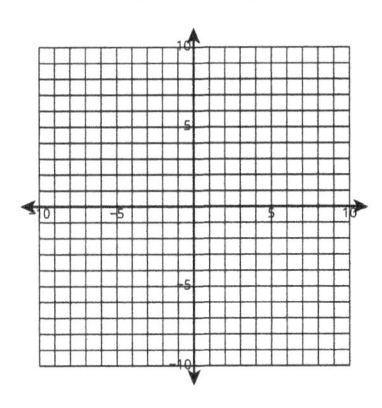

10. Solve the system of two linear inequalities graphically.

$$\begin{cases} y < 6 \\ y \geq -2 \end{cases}$$

  a. Graph the solution set of the **first** linear inequality.

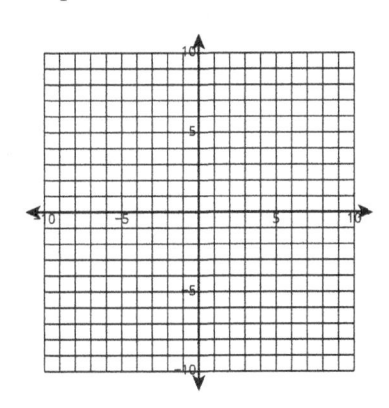

  b. Graph the solution set of the **second** linear inequality.

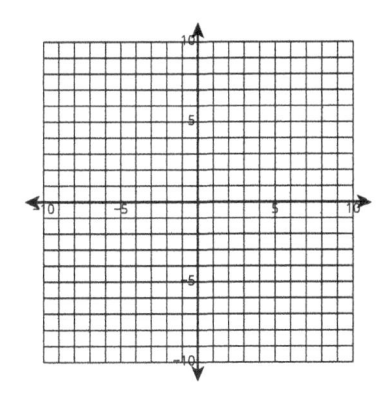

c. Find the region with points that satisfy both inequalities.

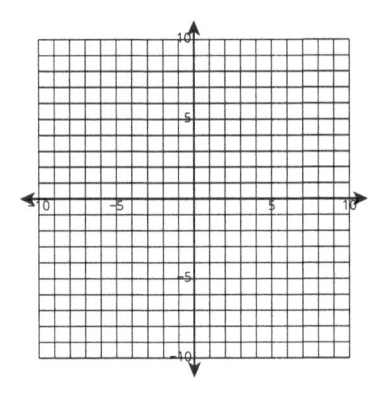

Applications

Solve.

11. Barbara's Bombtastic Bakery sells cookie bouquets where the price depends on the arrangement. Each completed bouquet arrangement needs to weigh less than 5 pounds for shipping purposes. The small cookies weigh 0.1 pounds and the large cookies weigh 0.3 pounds. The flower pot and Styrofoam weigh 1.2 pounds. The cost of each arrangement needs to be less than $30. The small cookies cost $1 each and the large cookies cost $2 each. (The cost of the flower pot and foam are included in the cookie prices.)

   a. Write two linear inequalities to describe the situation. Use the variable x to represent the number of small cookies and the variable y to represent the number of large cookies in a bouquet.

   b. Graph the two linear inequalities on the same coordinate plane.

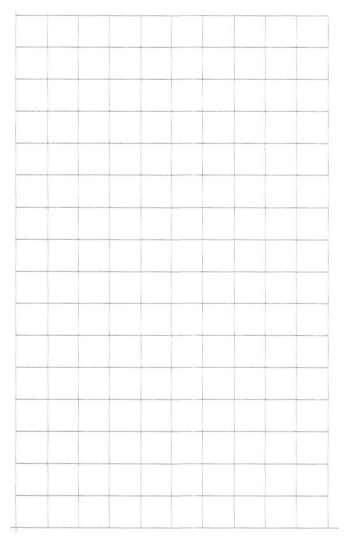

c. Describe the solution set for the situation.

d. Do any of the values in the solution set not make sense in the context of the problem? Explain why or why not.

Writing & Thinking

12. Explain in your own words how to test to determine which side of the graph of an inequality should be shaded.

Chapter 4
Algebraic Pathways: Exponents and Polynomials

4.1 Exponents

4.2 Scientific Notation

4.3 Modeling with Exponential Functions

4.4 Addition and Subtraction with Polynomials

4.5 Multiplication with Polynomials

4.1 Exponents

_____ can be used to indicate repeated multiplication by the same number (called the _____).

Product Rule for Exponents

If a is a nonzero real number and m and n are integers, then

$$\underline{\hspace{3cm}}.$$

In words, to multiply powers with the same base, _____.

The Exponent 0

If a is a nonzero real number, then

$$\underline{\hspace{2cm}}.$$

Note: The expression 0^0 is _____.

Quotient Rule for Exponents

If a is a nonzero real number and m and n are integers, then

$$\underline{\hspace{2cm}}.$$

In words, to divide two powers with the same base, _____ _____.

Rule for Negative Exponents

If a is a nonzero real number and n is an integer, then

$$\underline{\hspace{2cm}}.$$

In words, _____.

Power Rules for Exponents

If a and b are nonzero real numbers and m and n are integers:

1. **Power Rule:** _____

   To raise a power to a power, _____.

2. **Power Rule for Products:** _____

   To raise a product to a power, _____.

3. **Power Rule for Quotients:** _____

   To raise a quotients to a power,
   _____.

▶ Watch and Work

Watch the video for the example in the software and follow along in the space provided.

Example 6: Two Approaches with Fractional Expressions and Negative Exponents

Simplify $\left(\dfrac{-3a^2}{b^3}\right)^{-2}$.

Solution

✎ Now You Try It!

Use the space provided to work out the solution to the next example.

Example A: Two Approaches with Fractional Expressions and Negative Exponents

Simplify:

$\left(\dfrac{x^6}{y^3}\right)^{-5}$

Solution

Summary of Properties and Rules for Exponents

If a and b are nonzero real numbers and m and n are integers:

1. The exponent 1: _____

2. The exponent 0: _____

3. Product rule: _____

4. Quotient rule: _____

5. Negative exponents: _____

The Power Rules

6. Power rule: _____

7. Power of a product: _____

8. Power of a quotient: _____

4.1 Exercises

Concept Check

True/False. Determine whether each statement is true or false. If a statement is false, explain how it can be changed so the statement is true. (**Note:** There may be more than one acceptable change.)

1. If a constant does not have an exponent written, it is assumed that the exponent is 0.

2. If a is a nonzero real number and n is an integer, then $a^{-n} = -a^n$.

3. Since the product rule is stated for integer exponents, the rule is also valid for 0 and negative exponents.

4. When using the quotient rule, you should subtract the smaller exponent from the larger exponent.

Practice

5. Simplify the expression using the properties of exponents. Expand any numerical portion of your answer and only include positive exponents.

$x^4 \cdot x^3$

6. Simplify the expression using the properties of exponents. Expand any numerical portion of your answer and only include positive exponents.

$-5 \cdot 4^3$

7. Simplify the expression using the properties of exponents. Expand any numerical portion of your answer and only include positive exponents.

9^0

8. Simplify the expression using the properties of exponents. Expand any numerical portion of your answer and only include positive exponents.

$(8x)^0$

9. Simplify the expression using the properties of exponents. Expand any numerical portion of your answer and only include positive exponents.

$\dfrac{x^9}{x}$

10. Simplify the expression using the properties of exponents. Expand any numerical portion of your answer and only include positive exponents.

$$\frac{3^7}{3^6}$$

11. Simplify the expression using the properties of exponents. Expand any numerical portion of your answer and only include positive exponents.

$-4x^{-9}$

12. Simplify the expression using the properties of exponents. Expand any numerical portion of your answer and only include positive exponents.

$$\frac{x^5 \cdot x^9}{x^5}$$

13. Simplify the expression using the properties of exponents. Expand any numerical portion of your answer and only include positive exponents.

$\left(2x^{-4}\right)^{-2}$

14. Simplify the expression using the properties of exponents. Expand any numerical portion of your answer and only include positive exponents.

$\left(\dfrac{x^6}{y}\right)^2$

15. Simplify the expression using the properties of exponents. Expand any numerical portion of your answer and only include positive exponents.

$\left(\dfrac{xy^{-2}}{3x^4y^2}\right)^4$

Applications

Solve.

16. Rylee wants to move all her files to a new hard drive that has 2^{12} GB of storage on it. She wants to designate the same amount of storage for each of 2^4 projects. How much storage should be assigned to each project? Write your answer as a power of two.

17. Trey is studying patterns in bacteria. For a positive test result in his experiment, bacteria must grow in population at a minimum rate of 3^2 in 2^4 hours. If the initial population of the bacteria is 35 and his final measurement after 24 hours is 3^8, should he mark the test as positive or negative?

4.2 Scientific Notation

> **Scientific Notation**
>
> If N is a decimal number, then in **scientific notation**
>
> _____

▶ Watch and Work

Watch the video for the example in the software and follow along in the space provided.

Example 3: Application: Scientific Notation

Light travels approximately 3×10^8 meters per second. How many meters per minute does light travel?

Solution

✏ Now You Try It!

Use the space provided to work out the solution to the next example.

Example A: Application: Scientific Notation

One mole, a value often used in physics and chemistry, equals 6.02×10^{23} particles for all substances. How many particles would be in 8 moles of carbon?

Solution

4.2 Exercises

Concept Check

True/False. Determine whether each statement is true or false. If a statement is false, explain how it can be changed so the statement is true. (**Note:** There may be more than one acceptable change.)

1. The exponent in the number 1.4×10^4 indicates that the decimal point should be moved 4 places to the right.

2. The exponent in the number 2.5×10^{-3} indicates that the decimal point should be moved 3 places to the right.

3. The number 3.53×10^5 is less than 8.72×10^{-4}.

4. The number 4000 written in scientific notation is 0.4×10^4.

Practice

5. Write 25,000 in scientific notation.

6. Write 0.00000272 in scientific notation.

7. Simplify the expression $\dfrac{0.068 \times 0.0043}{0.34 \times 0.043}$ using scientific notation and express your answer in scientific notation.

8. Simplify the expression $\dfrac{4.9 \times 0.005 \times 60}{15 \times 0.0025 \times 800}$ using scientific notation and express your answer in scientific notation.

Applications

Solve.

9. An atom of gold has a mass of approximately 3.25×10^{-22} grams. What is the mass of 1000 atoms of gold? Express your answer in scientific notation.

10. A molecule of table salt weighs approximately 9.704×10^{-23} grams. What would be the weight of 560,000 molecules of table salt? Express your answer in scientific notation.

4.3 Modeling with Exponential Functions

Linear Functions

A **linear function** has the form

$$\underline{\hspace{5cm}},$$

where m and b are real numbers, and m represents the _____ .

Exponential Functions

An **exponential function** has the form

$$\underline{\hspace{4cm}},$$

where $b > 0$ and $b \neq 1$.

The value b is known as the _____ and the exponent x can take on the value of any real number.

Exponential Growth

Exponential growth is a specific form of an exponential function that results in growth, or an _____, over time. Exponential growth is modeled by

$$\underline{\hspace{4cm}},$$

where a is the _____ , $b > 1$, and a and x are any real numbers.

Exponential Decay

Exponential decay is a specific form of an exponential function that results in decay, or a _____, over time. Exponential decay is modeled by

$$\underline{\hspace{4cm}},$$

where a is the _____ , $0 < b < 1$, and x is any real number.

The Number e

The number e is an _____ number and is defined to be

$$e = \underline{}$$

We call e the _____.

Continually Compounded Interest

Continuously compounded interest is a special case of exponential growth. It is modeled by the formula

$$\underline{},$$

where

A is the _____,

P is the _____,

r is the _____, and

t is the _____.

▶ Watch and Work

Watch the video for the example in the software and follow along in the space provided.

Example 5: Application: Continuously Compounded Interest

An investment firm offers continuously compounded interest if $10,000 is invested for 5 years. The rate offered is 0.9%. What will be the value of the account after 5 years?

Solution

✏️ Now You Try It!

Use the space provided to work out the solution to the next example.

Example A: Application: Continuously Compounded Interest

An investment firm offers continuously compounded interest if $15,000 is invested for 10 years. The rate offered is 2.2%. What will be the value of the account after 10 years?

Solution

4.3 Exercises

Concept Check

True/False. Determine whether each statement is true or false. If a statement is false, explain how it can be changed so the statement is true. (**Note:** There may be more than one acceptable change.)

1. Exponential functions increase quickly at first and then grow very slowly.

2. For all exponential functions $f(x) = b^x$, $b < 0$.

3. The function $f(x) = 5^x$ is an example of exponential growth.

4. Continuously compounding interest compounds once a month.

Practice

5. Determine whether the data set is best described by a linear model or an exponential model.

Time	Amount
0	2
1	4
2	8
3	16
4	32
5	64
6	128
7	256

Applications

Solve.

6. The first leaves of autumn have fallen! On day 1, there were 3 leaves on the ground, and the number of leaves on the ground has been tripling with each passing day. This can be modeled by the function $f(x) = 3^x$. Graph the function.

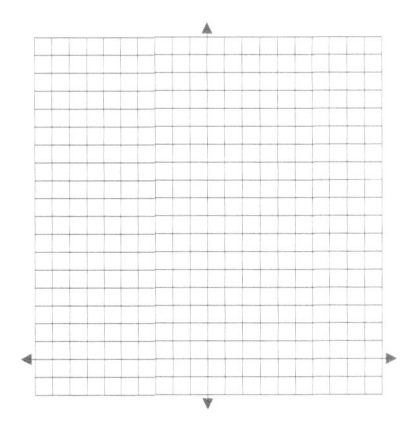

7. Growing linearly, the balance owed on your credit card doubles from $600 to $1200 in 6 months. If the balance were growing according to the exponential function $f(x) = 600(1 + 0.122)^x$ where x represents the number of months, what would the balance be after 6 months? Round your answer to the nearest cent.

8. If $7525 is invested at a rate of 7% compounded continuously, what will be the balance after 24 years? Round your answer to two decimal places.

9. Find the amount of money that will be accumulated in a savings account if $1750 is invested at 8.0% for 22 years and the interest is compounded continuously. Round your answer to two decimal places.

10. A tutoring service posted a new advertisement that offered a 20% discount on the first tutor session if the ad was mentioned. The table provides the number of callers who mentioned the advertisement during each of the first six days after the advertisement started running.

Days after Advertisement was Placed	Number of Phone Calls
1	1
2	3
3	5
4	10
5	15
6	26

Using the first column as the independent variable x and the second column as the dependent variable y, use a graphing calculator to compute the exponential model of the data.

4.4 Addition and Subtraction with Polynomials

A **monomial** is a term that has _____, and its variables have _____.

Monomial

A **monomial in x** is an expression of the form

$$\underline{}$$

where n is _____ and k is _____.

n is called _____ of the monomial, and k is _____.

The **degree of a monomial in more than one variable** is _____.

Polynomial

A **polynomial** is a monomial or _____.

The **degree of a polynomial** is _____.

The _____ is called the **leading coefficient**.

Classification of Polynomials

Term	Description	Example
Monomial:	polynomial with _____	_____
Binomial:	polynomial with _____	_____
Trinomial:	polynomial with _____	_____

No special name is given to polynomials with more than three terms. They are referred to simply as polynomials. Of course, monomials, binomials, and trinomials can be referred to as polynomials as well.

The **sum** of two or more polynomials can be found by _____.

4.4 Addition and Subtraction with Polynomials

▶ Watch and Work

Watch the video for the example in the software and follow along in the space provided.

Example 2: Adding Polynomials in a Vertical Format

Write the sum $(5x^3 - 9x^2 - 10x + 12) + (3x^3 + 6x^2 - 7)$ in a vertical format and evaluate.

Solution

✏ Now You Try It!

Use the space provided to work out the solution to the next example.

Example A: Adding Polynomials

Add

$(7x^3 + 2x^2 + 9)$
$\quad + (-3x^3 - 5x^2 + 7x - 7)$

Solution

To find the **difference** of two polynomials, either

a. _____ , or equivalently,

b. use the distributive property and

_____ .

4.4 Exercises

Concept Check

True/False. Determine whether each statement is true or false. If a statement is false, explain how it can be changed so the statement is true. (**Note:** There may be more than one acceptable change.)

1. A nonzero constant is a monomial with no degree.

2. A monomial is a polynomial with one term.

3. When subtracting one polynomial from another polynomial, only the first term of the polynomial is subtracted.

4. The terms $6a^2$ and $7a^2$ are not like terms because they don't have the same coefficient.

Practice

5. Determine if the following expression is a polynomial.

$x^3 + 8x + 6x^{-5}$

6. Consider the following expression:

$-7x^4 + 9$

   a. The given expression is a polynomial. Determine if the polynomial is a monomial, binomial, or trinomial.

   b. Determine the degree and the leading coefficient of the polynomial.

7. Perform the indicated operation by removing the parentheses and combining like terms.

$(4x - 6) + (-2x^2 + 7x - 6)$

8. Perform the indicated operation by removing the parentheses and combining like terms.

$(-2x^2 + 7x - 6) + (4x^2 - 6x - 7)$

9. Perform the indicated operation by removing the parentheses and combining like terms.

$(-2x + 7) - (-6x + 4)$

4.4 Addition and Subtraction with Polynomials

10. Perform the indicated operation by removing the parentheses and combining the like terms.

$(-8x^2 + 2x + 7) - (3x^2 - 6x - 6)$

11. Consider the function:

$f(x) = 3x^2 + 3x + 1$

    a. Find the value of $f(3)$.

    b. Find the value of $f(a)$.

Applications

Solve.

12. Chelsea is competing in a double-elimination softball tournament, which means that a team is eliminated once they lose two games. If the winning team goes undefeated, the total number of games played will be $G(t) = 4t + 2$, where t is the number of teams participating in the tournament.

    a. Identify the degree of the polynomial.

    b. Determine how many games will be played if 5 teams are participating.

13. The number of protozoa in a biology laboratory experiment is given by the polynomial function $p(t) = 0.02t^4 + 0.4t^3 + 6t^2$, where p is the number of protozoa after t hours.

    a. Determine the number of protozoa after 5 hours. Round your answer to the nearest whole number.

    b. How many protozoa are present after 2 days? Round your answer to the nearest whole number.

Writing & Thinking

14. Explain, in your own words, how to subtract one polynomial from another.

4.5 Multiplication with Polynomials

The distributive property is applied by _____

▶ Watch and Work

Watch the video for the example in the software and follow along in the space provided.

Example 7: Using the FOIL Method to Multiply Binomials

Multiply: $(2x - 3)(3x - 5)$

Solution

✏ Now You Try It!

Use the space provided to work out the solution to the next example.

Example A: Using the FOIL Method to Multiply Binomials

Multiply: $(4x - 5)(2x - 2)$

Solution

Difference of Two Squares

_____

Squares of Binomials (Perfect Square Trinomials)

_____ Square of a binomial sum

_____ Square of a binomial difference

170 Chapter 4 Algebraic Pathways: Exponents and Polynomials

4.5 Exercises

Concept Check

True/False. Determine whether each statement is true or false. If a statement is false, explain how it can be changed so the statement is true. (**Note:** There may be more than one acceptable change.)

1. The distributive property can only be used to multiply a monomial and a polynomial.

2. The FOIL method is a way to remember one specific order that the distributive property can be applied.

3. When two binomials are in the form of the sum and difference of the same term, the product will be a trinomial.

4. When the two binomials being multiplied together are the same, the product will be a trinomial.

Practice

5. Multiply the polynomial by the monomial using the distributive property and/or the product rule of exponents.

    $(-3)(2x^2 - 3x - 1)$

6. Multiply the polynomials using the distributive property and combine like terms.

    $(x+9)(x+1)$

7. Multiply the polynomials using the distributive property and combine like terms.

    $(x+4)(x^2 - x - 2)$

8. Multiply the binomials using the FOIL method. Combine like terms.

    $(x+1)(x+4)$

9. Multiply the binomials using the FOIL method. Combine like terms.

$(2x+2)(3x-3)$

10. Find the product of the binomials using the appropriate special product (difference of two squares, square of a binomial sum, or square of a binomial difference).

$(x+9)(x-9)$

11. Find the product of the binomials using the appropriate special product (difference of two squares, square of a binomial sum, or square of a binomial difference).

$(9x+1)(9x-1)$

12. Find the product of the binomials using the appropriate special product (difference of two squares, square of a binomial sum, or square of a binomial difference).

$(x+9)^2$

13. Find the product of the binomials using the appropriate special product (difference of two squares, square of a binomial sum, or square of a binomial difference).

$(9x+1)^2$

Applications

Solve.

14. A graphic artist is designing a poster to advertise an upcoming event. The only restrictions regarding the poster size is that it must have a length of $5x$ inches and a width of $7x+2$ inches. Find a simplified expression for the area of the poster.

15. A rectangle is $(x+6)$ ft by $(x+3)$ ft. If a square of length x ft on a side is cut from the rectangle, represent the remaining area in the form of a polynomial function $A(x)$.

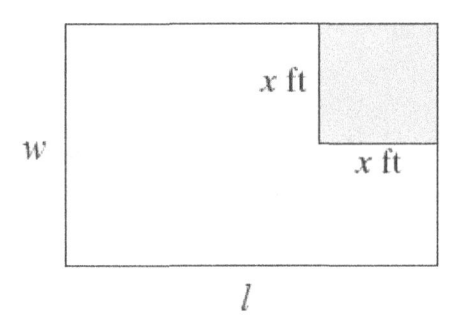

Writing & Thinking

16. Show how the distributive property can be used to find the product
$$\begin{array}{r} 75 \\ \times\ 93 \\ \hline \end{array}$$

(**Hint:** $75 = 70 + 5$ and $93 = 90 + 3$)

Chapter 5
Algebraic Pathways: Factoring and Solving Quadratic Equations

5.1 GCF and an Introduction to Factoring Polynomials

5.2 Factoring Trinomials

5.3 Special Factoring Techniques and General Guidelines for Factoring

5.4 Solving Quadratic Equations by Factoring

5.5 Operations with Radicals

5.6 Solving Quadratic Equations by the Square Root Property and the Quadratic Formula

5.7 Applications of Quadratic Equations

5.8 Graphing Quadratic Functions

5.1 GCF and an Introduction to Factoring Polynomials

The **greatest common factor** (**GCF**) of two or more integers is the

_____ .

Procedure for Finding the GCF of a Set of Terms

1. Find the prime factorization of _____ .

2. List all the factors that are _____ .

3. Raise each common factor to the

   _____ .

4. Multiply these powers to _____ .

Note: If there is no common prime factor or variable, then _____ .

Factoring Out the GCF

1. Find the GCF of the _____ .

2. Divide this monomial factor into

   _____ .

The product of the GCF and this new polynomial factor is _____

_____ .

▶ Watch and Work

Watch the video for the example in the software and follow along in the space provided.

Example 8: Factoring Polynomials by Grouping

Factor $xy + 5x + 3y + 15$ by grouping.

Solution

✏️ Now You Try It!

Use the space provided to work out the solution to the next example.

Example A: Factoring Polynomials by Grouping

Factor $x^2 + xy + x + y$ by grouping

Solution

5.1 Exercises

Concept Check

True/False. Determine whether each statement is true or false. If a statement is false, explain how it can be changed so the statement is true. (**Note:** There may be more than one acceptable change.)

1. When finding the GCF of a polynomial, you need to consider only the coefficients.

2. An expression is factored completely if none of its factors can be factored.

3. One way to find the GCF of a set of numbers is to use the prime factorization of each number.

4. Binomials cannot be factored out of algebraic expressions.

Practice

5. Find the GCF (greatest common factor) of the following terms.

   $\{17, 30, 60\}$

6. Find the GCF (greatest common factor) of the following terms.

   $\{90, 30xy^2, 60xy\}$

7. Divide the polynomial by the monomial denominator by writing the fraction as the sum (or difference) of fractions. Simplify your answer, if possible.

$$\frac{6x - 6}{-3x}$$

8. Divide the polynomial by the monomial denominator by writing the fraction as the sum (or difference) of fractions. Simplify your answer, if possible.

$$\frac{4x^3 - 3x + 7}{x^3}$$

9. Factor the given polynomial by finding the greatest common monomial factor (or the negative of the greatest common monomial factor) and rewrite the expression.

$35x^2y - 7y^2$

10. Factor the given polynomial by finding the greatest common monomial factor (or the negative of the greatest common monomial factor) and rewrite the expression.

$25x + 15x^2 + 25$

11. Factor the given polynomial by finding the greatest common monomial factor (or the negative of the greatest common monomial factor) and rewrite the expression.

$15x^2y + 33x^2y^2 + 3x^2y^3$

12. Completely factor the expression by grouping, if possible.

$d^2 + 6d - 7d - 42$

13. Completely factor the expression by grouping, if possible.

$3y^2 - 12 + y^2x + 3x$

14. Completely factor the expression by grouping, if possible.

$5ac - 8bd + bc - 40ad$

Applications

Solve.

15. A circus performer is shot vertically into the air with an initial velocity of 48 feet per second. The height of the performer above the ground in feet can be described by the polynomial $48x16x^2$ after x seconds.

   a. Find the height of the circus performer after 2 seconds.

   b. Factor the polynomial $48x16x^2$.

   c. Use the factored form of the polynomial from Part b. to find the height of the circus performer after 2 seconds.

   d. Are the answers from Parts a. and c. the same? Explain why or why not.

Writing & Thinking

16. Explain why the GCF of $-3x^2 + 3$ is 3 and not -3.

5.2 Factoring Trinomials

To factor a trinomial with leading coefficient 1, find _____

_____ . (If these factors do not exist, the trinomial is ___

_____)

▶ Watch and Work

Watch the video for the example in the software and follow along in the space provided.

Example 1: Factoring Trinomials with Leading Coefficient 1

Factor $x^2 - 10x + 16$.

Solution

✏ Now You Try It!

Use the space provided to work out the solution to the next example.

Example A: Factoring Trinomials with Leading Coefficients of 1

Factor:

$x^2 + 10x + 21$

Solution

Guidelines for the Trial-and-Error Method

1. If the sign of the constant term is positive (+), the signs in
_____ .

2. If the sign of the constant term is negative (−), the signs in
_____ .

Analysis of Factoring by the *ac*-Method

General Method

$$ax^2 + bx + c$$

Step 1:

Multiply _____ .

Step 2:

Find two integers whose product is *ac* and _____
_____ .

Step 3:

Rewrite the middle term (bx) using the
_____ .

Step 4:

Factor by grouping the
_____ .

Step 5:

Factor out the common binomial factor. This will give _____
_____ .

Example

$$2x^2 + 9x + 10$$

Step 1:

    Multiply _____ .

Step 2:

    Find two integers whose product is 20 and _____

    _____

Step 3:

    Rewrite the _____

    _____

Step 4:

    Factor by grouping the

    _____ .

    _____

Step 5:

    Factor out the

    _____ ,

    _____

5.2 Exercises

Concept Check

True/False. Determine whether each statement is true or false. If a statement is false, explain how it can be changed so the statement is true. (**Note:** There may be more than one acceptable change.)

1. In a trinomial such as $x^2 - 5x + 4$, one would need to find two factors of 4 whose sum is negative 5.

2. A trinomial is factorable if the middle term is the difference of the inner and outer products of two binomials.

3. The first step in the *ac*-method of factoring is to rewrite the middle term.

4. Factoring can be checked by multiplying the factors and verifying that the product matches the original polynomial.

Practice

5. Factor the given trinomial, if possible.

    $x^2 - 10x + 24$

6. Factor the given trinomial, if possible.

    $2x^2 - 20x + 48$

7. Completely factor the trinomial, if possible.

    $5x^2 - 17x + 14$

8. Completely factor the trinomial, if possible.

    $9x^2 + 30x - 24$

9. Completely factor the trinomial, if possible.

$p^2 + 15p + 15$

Writing & Thinking

10. Discuss, in your own words, how the sign of the constant term determines what signs will be used in the factors when factoring trinomials.

11. It is true that $2x^2 + 10x + 12 = (2x + 6)(x + 2) = (2x + 4)(x + 3)$. Explain how the trinomial can be factored in two ways. Is there some kind of error?

5.3 Special Factoring Techniques and General Guidelines for Factoring

Difference of Two Squares

Consider the polynomial $x^2 - 25$. By recognizing this expression as the **difference of two squares**, we can go directly to the factors.

$x^2 - 25 = $ _____

Sum of Two Squares

The **sum of two squares** is an expression of the form

_____ .

In a perfect square trinomial, both the first and last terms of the trinomial must be perfect squares. If the first term is of the form _____ and the last term is of the form _____, then the middle term must be of the form _____ or _____ .

▶ Watch and Work

Watch the video for the example in the software and follow along in the space provided.

Example 3: Factoring Perfect Square Trinomials

Factor completely.

a. $z^2 - 12z + 36$

b. $4y^2 + 12y + 9$

c. $2x^3 - 8x^2y + 8xy^2$

d. $(x^2 + 6x + 9) - y^2$

Solution

5.3 Special Factoring Techniques and General Guidelines for Factoring

✏ Now You Try It!

Use the space provided to work out the solution to the next example.

Example A: Factoring Perfect Square Trinomials

Factor completely

a. $z^2 + 40z + 400$

b. $y^2 - 14y + 49$

c. $3x^2 z - 18xyz + 27y^2 z$

d. $(y^2 + 8y + 16) - z^2$

Solution

$$(x + a)(x^2 - ax + a^2) = x^3 + a^3$$

$$(x - a)(x^2 + ax + a^2) = x^3 - a^3$$

5.3 Exercises

Concept Check

True/False. Determine whether each statement is true or false. If a statement is false, explain how it can be changed so the statement is true. (**Note:** There may be more than one acceptable change.)

1. The expression $x^2 + 20x + 100$ is a perfect square trinomial.

2. When factoring polynomials, always look for a common monomial factor first.

3. The sum of two squares, $(x^2 + a^2)$, is factorable.

4. Sixty-four is a perfect square and a perfect cube.

Practice

5. Completely factor the given polynomial, if possible.

    $9y^2 - 4x^2$

6. Completely factor the polynomial, if possible.

    $9x^2 + 25y^2$

7. Completely factor the polynomial, if possible.

    $25x^2 + 10x + 1$

8. Completely factor the polynomial, if possible.

    $9 - 30x + 25x^2$

5.3 Special Factoring Techniques and General Guidelines for Factoring

9. Completely factor the given polynomial, if possible.

$27y^3 + 64x^3$

10. Completely factor the given polynomial, if possible.

$27y^3 - 64$

11. Completely factor the given polynomial, if possible.

$8x^2 - x - 7$

Writing & Thinking

12. Compound interest is interest earned on interest. If a principal P is invested and compounded annually (once a year) at a rate of r, then the amount, A_1 accumulated in one year is $A_1 = P + Pr$.

In factored form, we have $A_1 = P + Pr = P(1 + r)$.

At the end of the second year the amount accumulated is $A_2 = (P + Pr) + (P + Pr)r$.

a. Write the expression for A_2 in factored form similar to that for A_1.

b. Write an expression for the amount accumulated in three years, A_3, in factored form.

c. Write an expression for A_n the amount accumulated in n years.

d. Use the formula you developed in Part c. and your calculator to find the amount accumulated if $10,000 is invested at 6% and compounded annually for 20 years.

5.4 Solving Quadratic Equations by Factoring

Quadratic Equations

Quadratic equations are equations that can be written in the form

_____ ; where a, b, and c are real numbers and $a \neq 0$.

Zero-Factor Property

If the product of two (or more) factors is 0, then _____

_____ That is, for real numbers a and b,

if $a \cdot b = 0$, then _____.

In general, a quadratic equation has _____ . In the special cases where the two factors are the same, there is only one solution. We call this a _____ .

▶ Watch and Work

Watch the video for the example in the software and follow along in the space provided.

Example 4: Solving Quadratic Equations by Factoring

Solve by factoring: $4x^2 - 4x = 24$

Solution

Now You Try It!

Use the space provided to work out the solution to the next example.

Example A: Solving Quadratic Equations by Factoring

Solve by factoring: $9x^2 - 27x = 36$

Solution

To Solve a Quadratic Equation by Factoring

1. Add or subtract terms as necessary so that _____
   and the equation is in the _____ where a, b,
   and c are real numbers and $a \neq 0$.

2. Factor completely. (If there are any fractional coefficients, _____
   _____.)

3. Set each nonconstant factor equal to
   _____ .

4. Check each solution, one at a time, in _____ .

5.4 Exercises

Concept Check

True/False. Determine whether each statement is true or false. If a statement is false, explain how it can be changed so the statement is true. (**Note:** There may be more than one acceptable change.)

1. When solving quadratic equations by factoring, it is important that all of the coefficients are integers.

2. The standard form for a quadratic equation is $ax^2 + bx = c$.

3. Not all quadratic equations can be solved by factoring.

4. All quadratic equations have two distinct solutions.

Practice

5. Solve the following equation.

$(9x + 8)(x - 10) = 0$

6. Solve the following equation by factoring.

$z^2 + 9z + 18 = 0$

7. Solve the following equation by factoring.

$x^3 - 35x = 2x^2$

8. Solve the following quadratic equation by factoring.

$\frac{3}{8}x^2 - 3x + \frac{45}{8} = 0$

9. Solve the quadratic equation by factoring.

$16x^2 - 24x + 9 = 0$

10. Solve the quadratic equation by factoring.

$3x^2 = -x + 10$

Applications

Solve.

11. A ball is dropped from the top of a building that is 784 feet high. The height of the ball above ground level is given by the polynomial function $h(t) = -16t^2 + 784$ where t is measured in seconds.

   a. How high is the ball after 3 seconds? 5 seconds?

   b. How far has the ball traveled in 3 seconds? 5 seconds?

   c. When will the ball hit the ground? Explain your reasoning in terms of factors.

12. A tennis ball is dropped from a building. The position of the ball after t seconds is given by the polynomial function $s(t) = -4.9t^2 + 490$, where s is the height in meters of the ball.

   a. Find $s(0)$. What does this value represent in the context of this problem?

   b. How high is the tennis ball 2 seconds after it has been dropped?

   c. How long before the tennis ball hits the ground?

Writing & Thinking

13. When solving equations by factoring, one side of the equation must be 0. Explain why this is so.

5.5 Operations with Radicals

Radical Terminology

The symbol $\sqrt{}$ is called _____.

The number under the _____.

The complete expression, such as $\sqrt{64}$, is called
_____.

Square Root

If a is a nonnegative real number, then

_____,

and

_____.

Properties of Square Roots

If a and b are **positive** real numbers, then

1. _____

2. _____

Simplest Form for Square Roots

A square root is considered to be in **simplest form** when _____.

_____ have the same index and radicand, or they can be simplified so that they have the same index and radicand.

5.5 Operations with Radicals

To Rationalize a Denominator Containing a Square Root

If the denominator contains a square root, _____

_____ .

▶ Watch and Work

Watch the video for the example in the software and follow along in the space provided.

Example 10: Rationalizing Denominators Containing Square Roots

Rationalize each denominator.

a. $\dfrac{5}{\sqrt{3}}$

b. $\dfrac{3}{7\sqrt{2}}$

c. $\dfrac{\sqrt{7}}{\sqrt{8}}$

Solution

✏ Now You Try It!

Use the space provided to work out the solution to the next example.

Example A: Rationalizing Denominators Containing Square Roots

Rationalize each denominator. Assume that all variables represent positive real numbers.

a. $\dfrac{8}{\sqrt{5}}$

b. $\dfrac{4}{9\sqrt{3}}$

c. $\dfrac{\sqrt{11}}{\sqrt{18}}$

Solution

5.5 Exercises

Concept Check

True/False. Determine whether each statement is true or false. If a statement is false, explain how it can be changed so the statement is true. (**Note:** There may be more than one acceptable change.)

1. If a number is squared and the principal square root of the result is found, that square root is always equal to the original number.

2. The simplest form of a radical expression can be found by using prime factorization.

3. The radicals $\sqrt{a}$ and $\sqrt[3]{a}$ are like radicals.

4. The fraction $\dfrac{\sqrt{2}}{3}$ is in simplest form.

Practice

5. Evaluate the radical expression.

$\sqrt{-25}$

6. Evaluate the radical expression.

$-\sqrt{4}$

7. Evaluate the radical expression.

$\sqrt{\dfrac{36}{49}}$

8. Estimate the radical by identifying which two consecutive integers it falls between.

$\sqrt{53}$

9. Simplify the following expression.

$\sqrt{20}$

10. Simplify the following expression.

$-\sqrt{\dfrac{25}{81}}$

11. Perform the indicated operations on the following radicals.

$4\sqrt{50} - 2\sqrt{2} + 4\sqrt{5}$

12. Multiply the following radicals and simplify your answer.

$\sqrt{6} \cdot \sqrt{3}$

13. Multiply and simplify the following radical expressions.

$\left(\sqrt{6} - 6\right)\left(\sqrt{6} - 2\right)$

14. Rationalize the denominator and simplify, if possible.

$\dfrac{\sqrt{21}}{\sqrt{6}}$

15. Rationalize the denominator and simplify, if possible.

$\dfrac{\sqrt{20}}{\sqrt{55}}$

Applications

Solve.

16. The following two formulas are used in electricity.

$\begin{aligned} P &= I^2 R \\ E^2 &= PR \end{aligned}$ where $\begin{cases} P &= \text{power (in watts)} \\ I &= \text{current (in amperes)} \\ E &= \text{voltage (in volts)} \\ R &= \text{resistance (in ohms, } \Omega\text{)} \end{cases}$

What is the current in amperes of a light bulb that produces 200 watts of power and has a 10 Ω resistance? Round your answer to two decimal places, if necessary.

17. A nut company is determining how to package their new type of party mix. The marketing department is experimenting with different-sized cans for the party mix packaging. The designers use the equation $r = \sqrt{\dfrac{V}{h\pi}}$ to determine the radius of the can for a certain height h and volume V. The company decides they want the can to have a volume of 972π cm^3. Find the radius of the can if the height is 4 cm. Keep your answers in simplified radical form.

Writing & Thinking

18. Under what conditions is the expression $\sqrt{a}$ not a real number?

5.6 Solving Quadratic Equations by the Square Root Property and the Quadratic Formula

Square Root Property

For a quadratic equation in the form $x^2 = c$, where c is nonnegative,

$$\underline{\hspace{6cm}}.$$

This can be written as $\underline{\hspace{3cm}}$.

Quadratic Formula

The solutions of the general quadratic equation $ax^2 + bx + c = 0$, where $a \neq 0$, are

$$\underline{\hspace{6cm}}.$$

▶ Watch and Work

Watch the video for the example in the software and follow along in the space provided.

Example 7: The Quadratic Formula

Use the quadratic formula to solve the following quadratic equation.

$2x^2 + x - 2 = 0$

Solution

5.6 Solving Quadratic Equations by the Square Root Property and the Quadratic Formula

✏ Now You Try It!

Use the space provided to work out the solution to the next example.

Example A: The Quadratic Formula

Use the quadratic formula to solve the following quadratic equation. $x^2 + 4x - 1 = 0$

Solution

5.6 Exercises

Concept Check

True/False. Determine whether each statement is true or false. If a statement is false, explain how it can be changed so the statement is true. (**Note:** There may be more than one acceptable change.)

1. Quadratic equations that are not easily solved using factoring might be solved by the square root method.

2. The quadratic formula will always work when solving quadratic equations.

3. When using the quadratic formula, if the discriminant is greater than zero, there are infinite solutions.

4. If the discriminant is less than zero, there is no real solution.

Practice

5. Solve the following quadratic equation by using the definition of a square root and write the solutions in simplified radical form.

    $9x^2 = 25$

6. Solve the following quadratic equation by using the definition of a square root and write the solutions in simplified radical form.

    $(x + 6)^2 = 64$

7. Use the discriminant, $b^2 - 4ac$, to determine the number of real solutions of the following quadratic equation. Then solve the quadratic equation using the formula $y = \dfrac{-b \pm \sqrt{b^2 - 4ac}}{2a}$.

$-2y^2 + 4y + 7 = 0$

8. Use the discriminant, $b^2 - 4ac$, to determine the number of real solutions of the following quadratic equation. Then solve the quadratic equation using the formula $y = \dfrac{-b \pm \sqrt{b^2 - 4ac}}{2a}$.

$2y^2 = -3$

9. Solve the following quadratic equation using the quadratic formula.

$x^2 - 2x - 2 = 0$

10. Solve the following quadratic equation using the quadratic formula.

$-7x^2 + 2 = -4x$

Applications

Solve.

11. An orange is thrown down from the top of a building that is 300 feet tall with an initial velocity of 6 feet per second. The distance of the object from the ground can be calculated using the equation $d = 300 - 6t - 16t^2$, where t is the time in seconds after the orange is thrown.

   a. On a balcony, a cup is sitting on a table located 100 feet from the ground. If the orange is thrown with the right aim to fall into the cup, how long will the orange fall? Round to the nearest hundredth. (Hint: The distance is 100 feet.)

   b. If the orange misses the cup and falls to the ground, how long will it take for the orange to splatter on the sidewalk? (Hint: What is the height of the orange when it hits the ground?)

5.6 Solving Quadratic Equations by the Square Root Property and the Quadratic Formula

c. Approximately how much longer would it take for the orange to fall to the sidewalk than it would for the orange to fall into the cup?

12. Merida is practicing archery with her recurve bow. Her target is the top of a 3-foot tall bale of hay that is 400 feet away. She aims at a 45° angle and shoots the arrow with an initial velocity of 140 feet per second. The height of the arrow can be described by $h = 99t - 16t^2 + 5$, where 99 is the vertical velocity of the arrow, h is the height of the arrow, and t is the time in seconds that passes after the arrow leaves the bow.

   a. Solve the equation $3 = 99t - 16t^2 + 5$ to determine the time in seconds when the height of the arrow will be 3 feet. Round your answer to the nearest hundredth.

   b. When shot at a 45° angle, the horizontal velocity of the arrow is also 99 feet per second. Use this velocity to determine how long will it take the arrow to reach the bale of hay? Round your answer to the nearest hundredth. (Hint: Use the $d = rt$ formula.)

   c. Did Merida hit the target, undershoot the target, or overshoot the target? (Hint: Compare the answers from Part **a.** and Part **b.**)

Writing & Thinking

13. Find an equation of the form $Ax^4 + Bx^2 + C = 0$ that has the four roots ± 2 and ± 3. Explain how you arrived at this equation.

Chapter 5   Algebraic Pathways: Factoring and Solving Quadratic Equations

5.7 Applications of Quadratic Equations

Attack Plan for Application Problems

1. Read the _____ .

2. Decide what is being asked for and
   _____ . It may
   help to organize a
   _____ .

3. Form and then solve an equation that _____ .

4. Check your solution
   _____ .

▶ Watch and Work

Watch the video for the example in the software and follow along in the space provided.

Example 2: Application: Solving Quadratic Equations

In an orange grove, there are 10 more trees in each row than there are rows. How many rows are there if there are 96 trees in the grove?

Solution

5.7 Applications of Quadratic Equations

✏️ Now You Try It!

Use the space provided to work out the solution to the next example.

Example A: Application: Solving Quadratic Equations

In a theater, there are four less seats in a row than there are rows. How many rows are there if there are 357 seats in the theater?

Solution

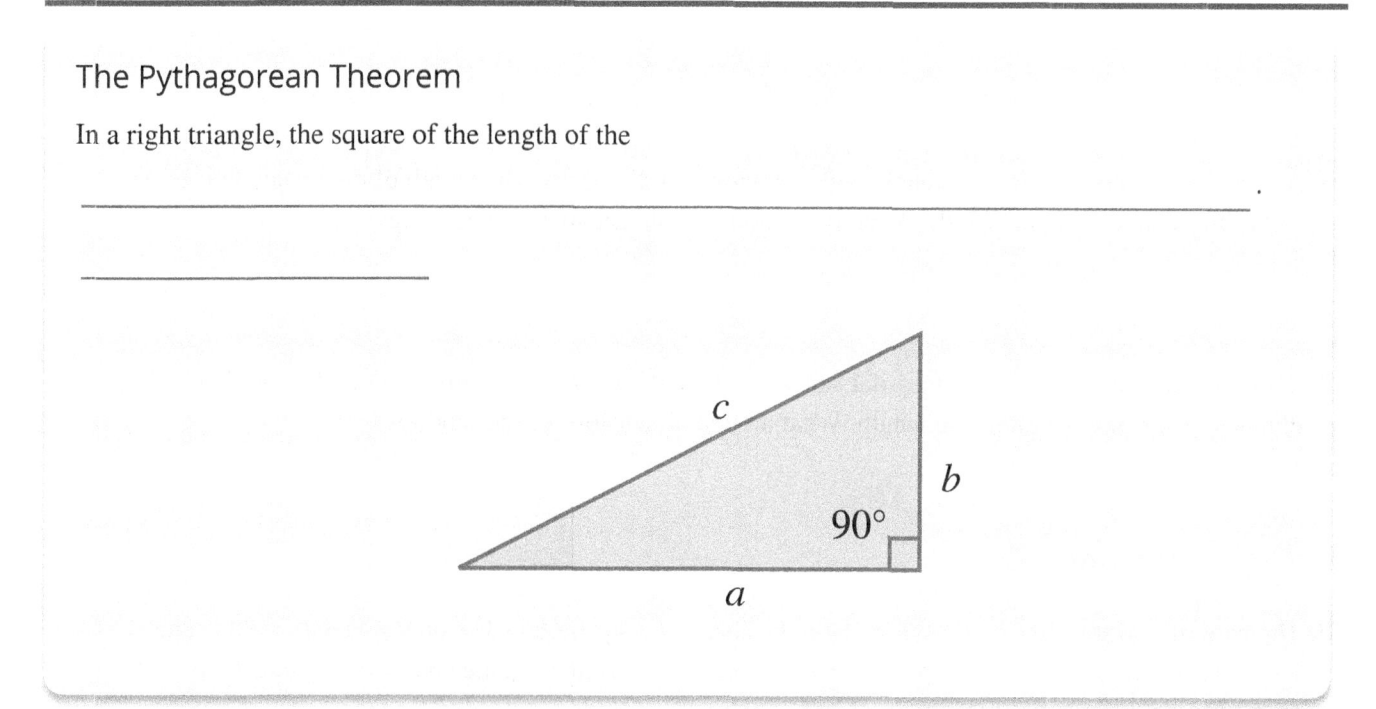

The Pythagorean Theorem

In a right triangle, the square of the length of the _____.

_____

5.7 Exercises

Concept Check

True/False. Determine whether each statement is true or false. If a statement is false, explain how it can be changed so the statement is true. (**Note:** There may be more than one acceptable change.)

1. The Pythagorean Theorem states that if the two legs of a right triangle are added, the sum will equal the hypotenuse.

2. The Pythagorean Theorem can be used with any triangle.

Applications

Solve.

3. One integer is 5 less than another. The sum of their squares is 125. Find the integers.

4. A theater can seat 320 people. The number of rows is 4 less than the number of seats in each row. How many rows of seats are there?

5. The perimeter of a rectangle is 48 feet. The area of the rectangle is 135 square feet. Find the dimensions of the rectangle.

6. An architect wants to draw a rectangle with a diagonal of 25 inches. The length of the rectangle is to be 3 inches more than triple the width. What dimensions should she make the rectangle?

7. Suppose you want to build a rectangular sandbox where the width is 2 feet more than the length and the diagonal is 4 feet longer than the length. What are the dimensions of the sandbox?

Writing & Thinking

8. The pattern in Kara's linoleum flooring is in the shape of a square 8 inches on a side with right triangles (with legs whose lengths are x inches) placed on each side of the original square so that a new larger square is formed. What is the area of the new square? Explain why you do not need to find the value of x.

9. Suppose that you are to solve an applied problem and the solution leads to a quadratic equation. You decide to use the quadratic formula to solve the equation. Explain what restrictions you must be aware of when you use the formula.

5.8 Graphing Quadratic Functions

Quadratic Function

A **quadratic function** is any function that can be written in the form

_____

where a, b, and c are real numbers and $a \neq 0$.

General Information on Quadratic Functions

For the **quadratic function** $y = ax^2 + bx + c$:

1. If $a > 0$, _____ .

2. If $a < 0$, _____ .

3. $x = -\dfrac{b}{2a}$ is the _____ .

4. The **vertex** (turning point) occurs _____ .

▶ Watch and Work

Watch the video for the example in the software and follow along in the space provided.

Example 1: Graphing a Quadratic Function (b = 0)

For the quadratic function $y = x^2 - 1$, find

  a. its vertex,

  b. its line of symmetry,

  c. its x-intercepts, and

  d. its y-intercept.

Plot a few specific points and graph the parabola.

Solution

✏️ Now You Try It!

Use the space provided to work out the solution to the next example.

Example A: Graphing a Quadratic Functions $(b = 0)$

For the following quadratic function $y = x^2 - 4$, find

a. its vertex

b. its line of symmetry

c. its *x*-intercepts

d. its *y*-intercept.

Plot a few specific points and graph the parabola.

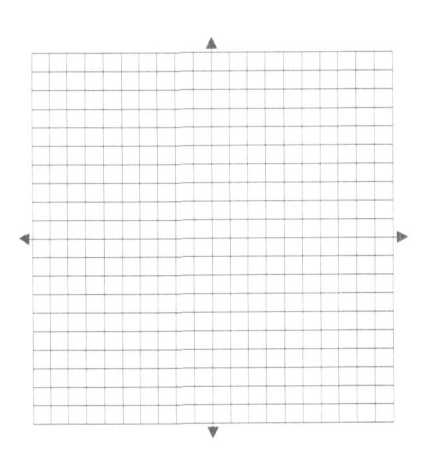

Solution

Minimum and Maximum Values

For a parabola with its equation in the form $y = a(x-h)^2 + k$,

1. If $a > 0$, then the parabola opens _____ .

2. If $a < 0$, then the parabola opens _____ .

5.8 Exercises

Concept Check

True/False. Determine whether each statement is true or false. If a statement is false, explain how it can be changed so the statement is true. (**Note:** There may be more than one acceptable change.)

1. The vertex of a vertical parabola is the lowest point on the parabola.

2. The maximum or minimum value of a quadratic function written in general form can be found by letting $x = -\dfrac{b}{2a}$ and solving for y.

3. Quadratic functions of the form $y = ax^2 + bx + c$ have a line of symmetry at $x = \dfrac{b}{2a}$

Practice

4. Consider the following quadratic function.

   $y = -x^2 - 10x - 30$

   a. Find the vertex of this function.

   b. Find the line of symmetry of this function.

c. Find the *x*-intercept(s), if any. Express your answer as ordered pair(s).

d. Find the *y*-intercept. Express your answer as an ordered pair.

e. Determine two points on the graph of the parabola other than the vertex and the *x*- and *y*-intercepts.

f. Graph the quadratic function.

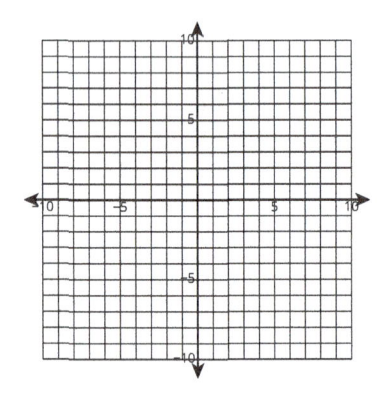

5. Consider the following quadratic function.

$$y = -x^2 + 4x + 1$$

a. Find the vertex of this function.

b. Find the line of symmetry of this function.

c. Find the *x*-intercept(s), if any. Express your answer as ordered pair(s).

d. Find the *y*-intercept. Express your answer as an ordered pair.

e. Determine two points on the graph of the parabola other than the vertex and the *x*- and *y*-intercepts.

f. Graph the quadratic function.

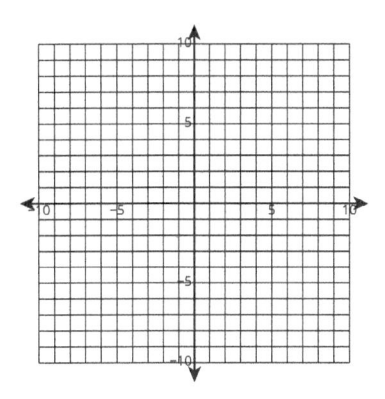

6. Consider the following quadratic function.

$y = x^2 - 2$

   a. Find the vertex of this function.

   b. Find the line of symmetry of this function.

   c. Find the *x*-intercept(s), if any. Express your answer as ordered pair(s).

   d. Find the *y*-intercept. Express your answer as an ordered pair.

   e. Determine two points on the graph of the parabola other than the vertex and the *x*- and *y*-intercepts.

f. Graph the quadratic function.

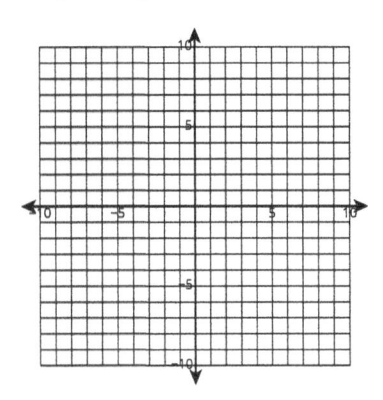

Applications

Solve.

7. A stone is projected vertically upward from a platform that is 13 ft high at a rate of 101 ft / sec. Use $h = -16t^2 + v_0 t + h_0$.

   a. Determine when the stone will reach its maximum height. (Round your answer to two decimal places.)

   b. Determine what the maximum height will be. (Round your answer to two decimal places.)

8. A store owner estimates that by charging x dollars each for a certain lamp, he can sell $50 - x$ lamps each week. What price will yield maximum revenue?

9. A contractor is to build a brick wall 6 feet hight to enclose a rectangular garden. Only three sides of the garden need to be enclosed becasue the fourth side is the building. The owner wants to enclode the maximum area but will only pay for 204 feet of wall.

   What dimensions should the contractor make the garden?

Writing & Thinking

10. Discuss the following features of the general quadratic function $y = ax^2 + bx + c$.

    a. What type of curve is its graph?

b. What is the value of x at its vertex?

c. What is the equation of the line of symmetry?

d. Does the graph always cross the x-axis? Explain.

Chapter 6
Geometric Pathways: Measurement & Geometry

6.1 US Measurements

6.2 The Metric System: Length and Area

6.3 The Metric System: Capacity and Weight

6.4 US and Metric Equivalents

6.5 Angles

6.6 Triangles

6.7 Perimeter and Area

6.8 Volume and Surface Area

6.9 Right Triangle Trigonometry

6.1 US Measurements

Using Multiplication and Division to Convert Measurements

1. **Multiply** to convert to _____. (There will be _____ _____.)

2. **Divide** to convert to _____. (There will be _____.)

Using Unit Fractions to Convert Measurements

1. The numerator should be in the _____.

2. The denominator should be in the _____.

▶ Watch and Work

Watch the video for the example in the software and follow along in the space provided.

Example 6: Application: Converting US Units of Measure

Determine how many seconds are in a 5-day work week assuming an 8 hr work day.

Solution

✏ Now You Try It!

Use the space provided to work out the solution to the next example.

Example A: Application: Converting US Units of Measure

How many fluid ounces are in 8 gallons of apple juice?

Solution

6.1 Exercises

Concept Check

True/False. Determine whether each statement is true or false. If a statement is false, explain how it can be changed so the statement is true. (**Note:** There may be more than one acceptable change.)

1. Capacity can be measured using ounces, quarts, and gallons.

2. One mile is equivalent to 2000 feet.

3. To convert from smaller units to larger units, division will be required.

4. Multiplication by a unit fraction does not change the value of the expressions being converted.

Practice

5. Convert the units of measure as indicated.

   135 in. = _____ ft

6. Convert the units of measure as indicated.

   34 lb = _____ oz

7. Convert the units of measure as indicated.

   49 qt = _____ pt

8. Convert the units of measure as indicated.

   375 sec = _____ min

Applications

Solve.

9. Sheer fabric costs $7.99 per yard. If it will take 35 feet of fabric to make drapes for the entire house, how much must you spend on fabric for the drapes, to the nearest cent?

10. While cleaning out the garage, Nelson discovers some containers of oil that need to be taken in for recycling. The containers hold 20 fluid ounces, 3 cups, and 1 quart, respectively. Find the total amount of oil ready to be recycled (in fluid ounces).

Writing & Thinking

11. Colby needs to find out how many yards are in one mile. What two sets of equivalent units would he need to make that determination?

6.2 The Metric System: Length and Area

Writing Metric Units of Measure

In the metric system,

1. A 0 is written to the left of the decimal point _____ . For example, _____ .

2. No commas are used in writing numbers. If a number has more than four digits (to the left or right of the decimal point), the digits are _____ .

   For example, _____ .

There are two basic methods of converting units of measurement in the metric system:

1. Multiplying by _____ .

2. Moving the _____ .

Using Unit Fractions to Convert Measures

1. The numerator should be in the _____ .

2. The denominator should be in the _____ .

▶ Watch and Work

Watch the video for the example in the software and follow along in the space provided.

Example 6: Converting Metric Units of Area

Convert each measurement using unit fractions.

   a. $5 \text{ cm}^2 = $ _____ mm^2

   b. $4600 \text{ mm}^2 = $ _____ m^2

Solution

✏️ Now You Try It!

Use the space provided to work out the solution to the next example.

Example A: Converting Metric Units of Area

Convert each measurement using unit fractions.

a. $86 m^2 = $ ____ cm^2

b. $0.06 mm^2 = $ ____ dm^2

Solution

6.2 Exercises

Concept Check

True/False. Determine whether each statement is true or false. If a statement is false, explain how it can be changed so the statement is true. (**Note:** There may be more than one acceptable change.)

1. To change from smaller units to larger units, multiplication must be used.

2. Units of length in the metric system are named by putting a prefix in front of the basic unit meter, for example, centimeter.

3. In metric units, a square that is 1 centimeter long on each side is said to have an area of 1 centimeter.

Practice

4. How many dekameters are in 203 cm?

5. Express 466.9 dam in centimeters.

6. Convert the following measurement.

   158,740,000,000 bytes = _____ gigabytes

7. Express 909 m² in square centimeters.

Applications

Solve.

8. A triangle has a base measuring 4 cm and a height measuring 16 mm. Determine the area of the triangle in cm².

9. A section of railroad track measuring 2.1 km in length needs to be replaced. Each railroad tie is 4 decimeters wide and they are to be spaced 0.8 m apart. How many railroad ties will be needed to complete this section of track?

Writing & Thinking

10. Compare and contrast ease of converting units in the US customary system and the metric system.

6.3 The Metric System: Capacity and Weight

In the metric system, capacity (liquid volume) is measured in _____ (abbreviated _____). A liter is the volume enclosed in a cube that is _____.

Mass is _____ in an object.

The basic unit of mass in the metric system is the _____ ,

▶ Watch and Work

Watch the video for the example in the software and follow along in the space provided.

Example 7: Converting Metric Units of Weight

Convert the units as indicated **a.** using a unit fraction and **b.** using a metric conversion line.

34 g = _____ mg

Solution

✏ Now You Try It!

Use the space provided to work out the solution to the next example.

Example A: Converting Metric Units of Weight

Convert 14.9 kg to grams using a unit fraction or a metric conversion line.

Solution

6.3 Exercises

Concept Check

True/False. Determine whether each statement is true or false. If a statement is false, explain how it can be changed so the statement is true. (**Note:** There may be more than one acceptable change.)

1. One milliliter is equivalent to one cubic centimeter.

2. Volume is measured in square units.

3. In 1 liter there are 100 milliliters.

4. A metric ton and a US customary ton are equal (a metric ton weighs about 2000 US pounds).

5. A dekagram contains 10 grams.

Practice

6. What metric unit would be the best choice to measure the mass of a drop of water?

7. What metric unit would be the best choice to measure the volume of liquid in a bottle of cough medicine?

8. Convert 0.785 L to milliliters using unit fractions. Do not round your answer.

9. Convert 49 000 mg to kilograms using unit fractions. Do not round your answer.

10. Convert 0.34 kg to grams using unit fractions. Do not round your answer.

Applications

Solve.

11. How many 5-mL doses of liquid medication can be given from a vial containing 3 deciliters?

12. One cup of flour is approximately 120 grams. How many cups of flour can you get out of a bag of flour weighing 2.4 kg?

Writing & Thinking

13. In the metric system, the common unit of capacity is the liter. Discuss how you would change from a measure of liters to milliliters.

226   Chapter 6   Geometric Pathways: Measurement & Geometry

6.4 US and Metric Equivalents

Temperature

US customary measure is in _____ .

Metric measure is in _____ .

Temperature Formulas

F = Fahrenheit temperature   and C = Celsius temperature

$F = $ _____ $C = $ _____

▶ Watch and Work

Watch the video for the example in the software and follow along in the space provided.

Example 5: Converting Units of Area

Convert each measurement, rounding to the nearest hundredth.

a. $40 \text{ yd}^2 = $ _____ m^2

b. $100 \text{ cm}^2 = $ _____ in.^2

c. 6 acres = _____ ha

d. 5 ha = _____ acres

Solution

✏️ Now You Try It!

Use the space provided to work out the solution to the next example.

Example A: Converting Units of Area

Convert each measurement, rounding to the nearest hundredth.

a. 53 in.^{2} = _____ cm^2

b. 50 m^2 = _____ ft^2

c. 16 acres = _____ ha

d. 3 ha = _____ acres

Solution

6.4 Exercises

Concept Check

True/False. Determine whether each statement is true or false. If a statement is false, explain how it can be changed so the statement is true. (**Note:** There may be more than one acceptable change.)

1. Water freezes at 32 degrees Celsius.

2. When converting between US customary and metric units, often the results will be approximations.

3. A 5K (km) run is longer than a 5 mile run.

4. One square meter covers more area than one square yard.

Practice

5. 30 °C = _____ °F. Round your answer to the nearest thousandth, if necessary.

6. Convert 150 yd to meters. Round your answer to the nearest thousandth, if necessary.

7. Convert 55 in.² to square centimeters. Round your answer to the nearest thousandth, if necessary.

8. What number of acres is equivalent to 800 ha? Round your answer to the nearest thousandth, if necessary.

9. What number of square yards is equivalent to 200 m²? Round your answer to the nearest thousandth, if necessary.

10. Convert 7 liters to quarts using unit fractions. Round your answer to the nearest thousandth, if necessary.

11. Convert 22 liters to gallons using unit fractions. Round your answer to the nearest thousandth, if necessary.

12. 5 lb = _____ kg. Round your answer to the nearest thousandth, if necessary.

13. 10 g = _____ oz. Round your answer to the nearest thousandth, if necessary.

Applications

Solve.

14. Allison buys a spool of thread for sewing. There are 6 yards of thread on the spool. She uses 3.5 meters. How much thread is left on the spool in meters? Round your answer to the nearest thousandth, if necessary.

15. Suppose it takes John 9 minutes to run 1 mile. How long would it take him to run 4 kilometers? Round your answer to the nearest minute.

Writing & Thinking

16. Most conversions between the US customary system of measure and metric system are not exact. Explain why this is true and give any exceptions.

6.5 Angles

Point, Line, Plane

Undefined Term	Representation	Discussion
Point	Point A $A \bullet$	A point is _____. Points are labeled with _____.
Line	Line l or line $\overleftrightarrow{AB}$ $l \leftarrow\!\bullet\!\!\!\underset{A}{}\!\!\!\!\!\underset{}{}\!\!\!\bullet\!\!\!\underset{B}{}\!\rightarrow$	A line has _____. Lines are labeled with _____.
Plane	Plane P	_____, represent planes. Planes are labeled with _____.

6.5　Angles　231

Ray and Angle

Term	Definition	Illustrations with Notation
Ray	A **ray** consists of _____ _____ .	Ray $\overrightarrow{PQ}$ with endpoint P
Angle	An **angle** consists of _____ _____ (The two rays are called the _____ and the endpoint is called _____ .)	$\angle AOB$ with vertex O

Angles Classified by Measure

Name	Measure	Illustrations with Notation
Acute	_____	$\angle A$ is an acute angle.
_____	_____	$\angle B$ is a right angle.

		![angle with vertex C]
		C
Straight		∠D is a straight angle. (The rays are in opposite directions.)

Complementary and Supplementary Angles

1. Two angles are _____ .

2. Two angles are _____ .

If two angles _____ , they are said to be
_____ (symbolized as ≅).

Two lines **intersect** if they have _____ . If two lines intersect,
_____ .

Vertical Angles

Vertical angles are _____ .

_____ .

Adjacent Angles

Two angles are **adjacent** if _____ .

6.5 Angles

Parallel Lines and Perpendicular Lines

Term	Definition	Illustrations with Notation
Parallel Lines	Two lines are **parallel** _____ _____	$\overleftrightarrow{PQ}$ is parallel to $\overleftrightarrow{RS}$ $\left(\overleftrightarrow{PQ} \parallel \overleftrightarrow{RS}\right)$
Perpendicular Lines	Two lines are **perpendicular** _____ _____	$\overleftrightarrow{PQ}$ is perpendicular to $\overleftrightarrow{RS}$ $\left(\overleftrightarrow{PQ} \perp \overleftrightarrow{RS}\right)$

A **transversal** is a line in a plane that _____

_____.

Parallel Lines and a Transversal

If two parallel lines are cut by a transversal, then the following two statements are true.

1. _____

2. _____

▶ Watch and Work

Watch the video for the example in the software and follow along in the space provided.

Example 8: Calculating Measures of Angles

In the figure below, lines k and l are parallel, t is a transversal, and $m\angle 1 = 50°$. Find the measures of the other 7 angles.

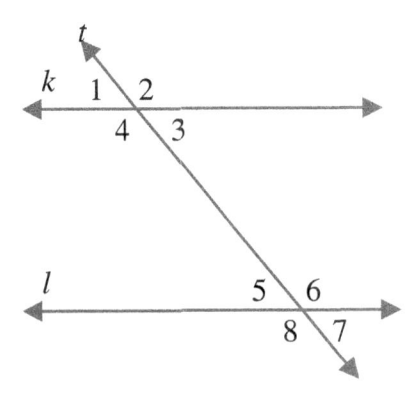

$m\angle 1 = 50°$

Solution

✏️ Now You Try It!

Use the space provided to work out the solution to the next example.

Example A: Calculating Measures of Angles

In the figure below, lines l and m are parallel, t is a transversal, and

$m\angle 2 = 80°$.

Find

$m\angle 4$,

$m\angle 5$,

and $m\angle 6$.

Solution

6.5 Exercises

Concept Check

True/False. Determine whether each statement is true or false. If a statement is false, explain how it can be changed so the statement is true. (**Note:** There may be more than one acceptable change.)

1. The sum of the measures of two complementary angles is equal to the measure of one right angle.

2. The sum of the measures of complementary angles is greater than the sum of the measures of supplementary angles.

3. Adjacent angles are two angles that share a side.

4. If two lines in a plane are not parallel, then they are perpendicular.

Practice

5. Classify the following angle.

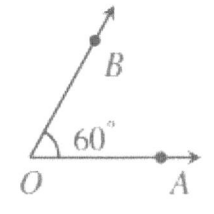

6. Classify the following angle.

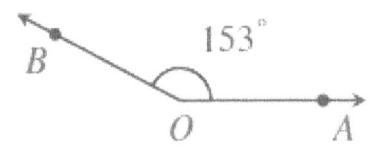

7. If ∠1 and ∠2 are complementary angles and m ∠1 = 82°, what is m ∠2?

8. If ∠1 and ∠2 are supplementary angles and m ∠1 = 163°, what is m ∠2?

9. The following figure shows three intersecting straight lines: x, y, and z, with m∠5 = 39° and m∠6 = 90°

   a. Find the measure of ∠1.

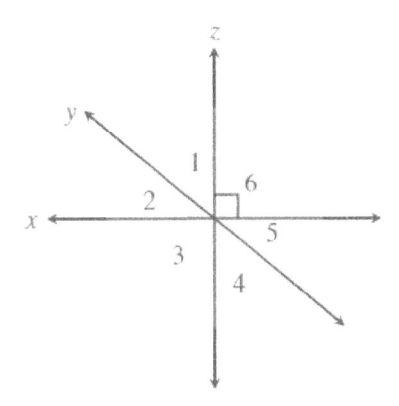

   b. Find the measure of ∠3.

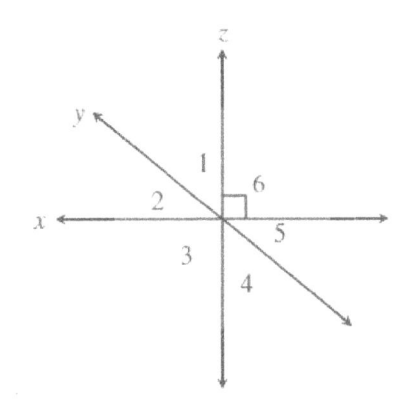

   c. Which two angles are complementary to ∠1?

10. The following figure shows two intersecting lines.

    a. If m∠1 = 122°, then what is m∠2

b. Name one pair of vertical angles.

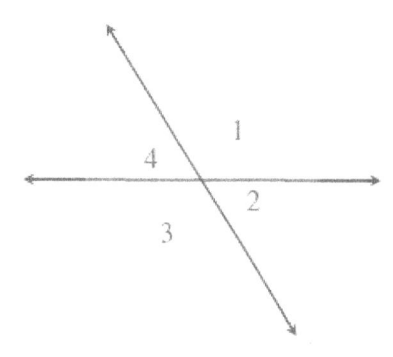

c. Name one pair of adjacent angles.

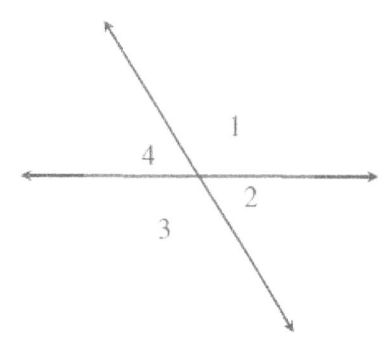

11. Consider the minute and hour hands on a clock. What type of angle is formed by the hands on a clock when it is exactly 6 o'clock?

Writing & Thinking

12. Explain, in your own words, the relationships between vertex, ray, angle, and line.

6.6 Triangles

A **line segment** consists of _____.

A **triangle** consists of _____
_____.

The line segments are called the _____ and the points are called the
_____.

Triangles are classified in two ways:

1. _____.

2. _____.

Triangles Classified by Sides
(**Note:** In the figures, sides with equal length are indicated by the same number of slash marks.)

Name	Property	Example
Scalene		$\triangle ABC$ is scalene since _____.
Isosceles		$\triangle PQR$ is isosceles since _____.
Equilateral		$\triangle XYZ$ is equilateral since _____.

Chapter 6 Geometric Pathways: Measurement & Geometry

Triangles Classified by Angles

Name	Property	Example
Acute		_____, so △ ABC is acute.
Right		_____, so △ PQR is a right triangle.
Obtuse		_____, so △ XYZ is an obtuse triangle.

Three Properties of Triangles

In a triangle, the following are true:

1. The sum of the measures _____ .

2. The sum of the lengths of
   _____ .

3. Longer sides are _____ .

Similar Triangles

1. In similar triangles, the **corresponding** _____ .

2. In similar triangles, the
   _____ .

Properties of Congruent Triangles

Two triangles are congruent if

1. _____ .

2. _____ .

Determining Congruent Triangles

1. **Side-Side-Side** (SSS)

   If two triangles are such that _____

   _____

2. **Side-Angle-Side** (SAS)

   If two triangles are such that the _____

   _____

   _____ .

3. **Angle-Side-Angle** (ASA)

   If two triangles are such that _____

   _____

▶ Watch and Work

Watch the video for the example in the software and follow along in the space provided.

Example 7: Determining Whether Triangles are Congruent

Determine whether triangles PQR and MNO are congruent.

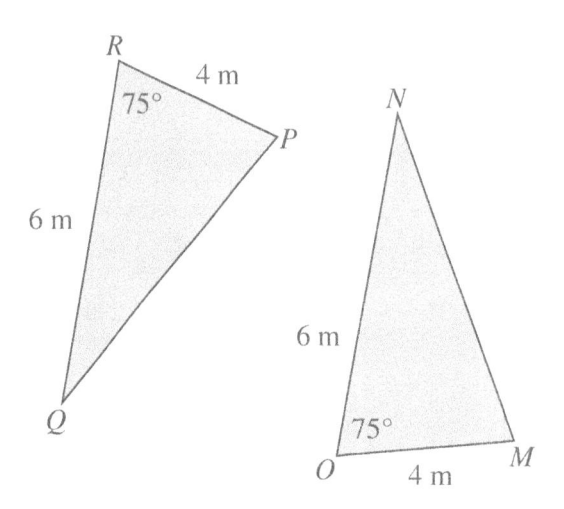

Solution

6.6   Triangles   243

✏️ Now You Try It!

Use the space provided to work out the solution to the next example.

Example A: Determining Whether Triangles are Congruent

Determine whether triangles *JKL* and *MNO* are congruent.

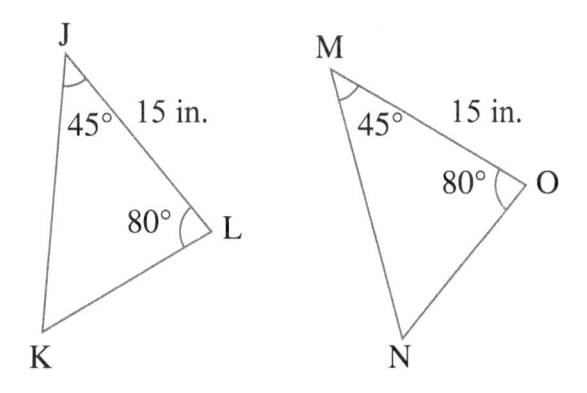

Solution

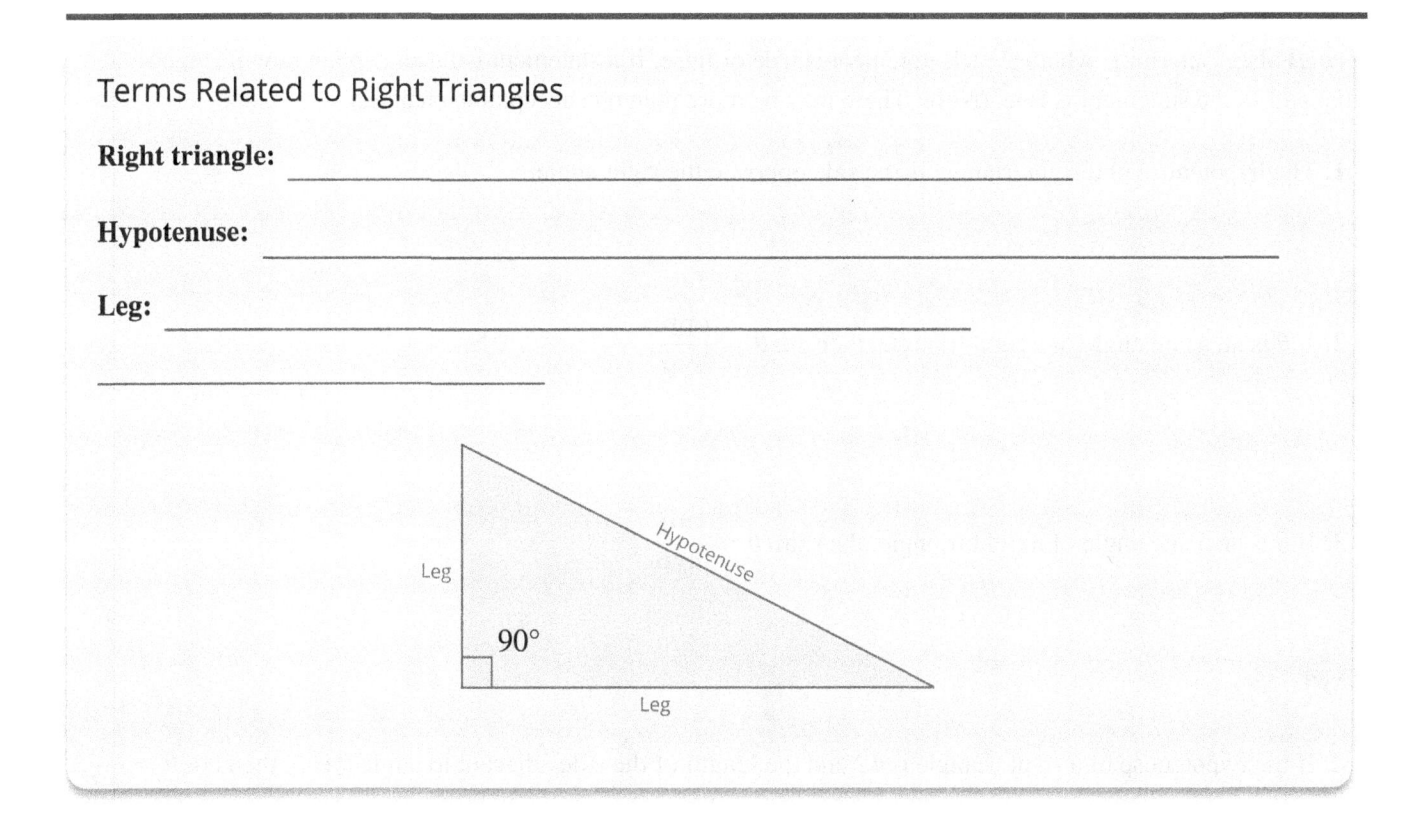

Terms Related to Right Triangles

Right triangle: _____

Hypotenuse: _____

Leg: _____

_____

The Pythagorean Theorem

In a right triangle, the _____

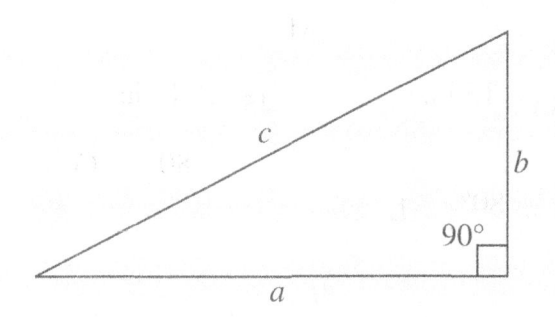

_____ .

_____ = _____ + _____

6.6 Exercises

Concept Check

True/False. Determine whether each statement is true or false. If a statement is false, explain how it can be changed so the statement is true. (**Note:** There may be more than one acceptable change.)

1. The hypotenuse of a right triangle is the side opposite the right angle.

2. If θ is an acute angle of a right triangle, then $\sin \theta = \dfrac{\text{opp}}{\text{hyp}}$.

3. If θ is an acute angle of a right triangle, then $\tan \theta = \dfrac{\text{adj}}{\text{hyp}}$.

4. If the hypotenuse of a right triangle is 12 and the length of the side adjacent to angle θ is 5, then $\cos \theta = \dfrac{5}{12}$.

Practice

5. Classify the following triangle as scalene, isosceles, or equilateral by the lengths of its three sides: 11 mm, 15 mm, 18 mm.

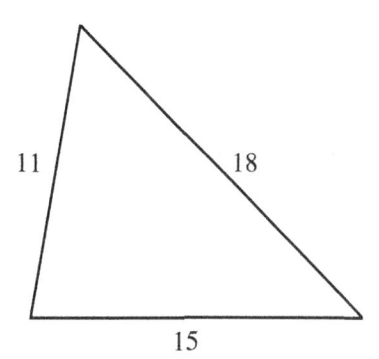

6. Suppose the lengths of the sides of △ABC are as shown in the figure. Is this possible? Explain your reasoning.

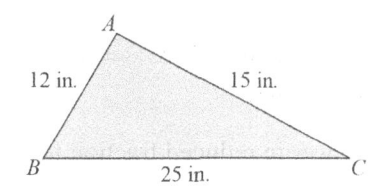

7. In the triangle shown, $m\angle X = 30°$ and $m\angle Y = 70°$.

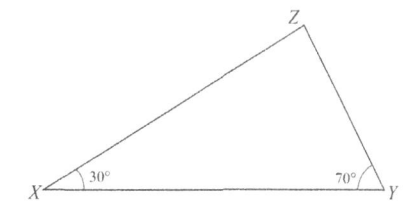

    a. What is $m\angle Z$?

    b. What kind of triangle is △XYZ?

    c. Which side is opposite $\angle X$?

d. Which sides include $\angle X$?

e. Is $\triangle XYZ$ a right triangle?

8. Determine whether this pair of triangles is similar. If the pair of triangles is similar, explain why and indicate the similarity by using the ~ symbol.

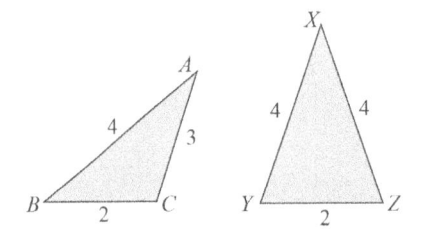

9. Consider the triangles $\triangle ABC$ and $\triangle DEF$ where $\triangle ABC \sim= \triangle DEF$.
(NOTE: these triangles are not drawn to scale)

   a. Find the value of x. Write your answer in reduced fraction form.

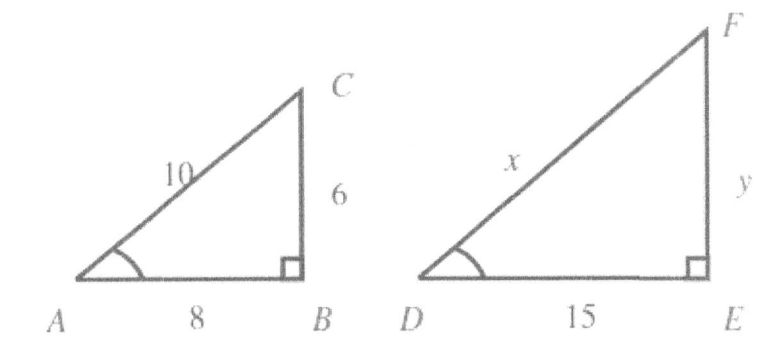

b. Find the value of *y*. Write your answer in reduced fraction form.

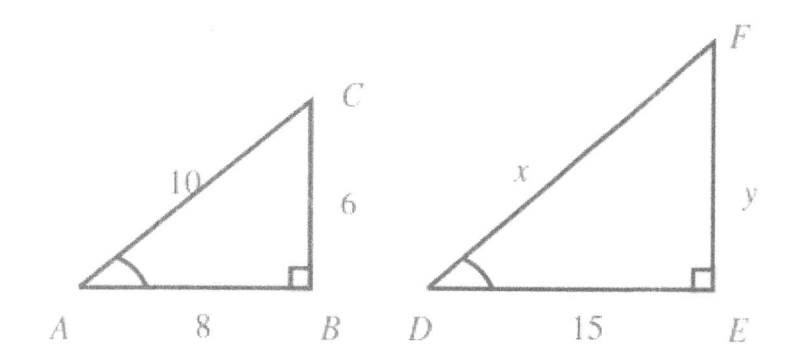

10. Determine whether this pair of triangles is congruent. If the pair of triangles is congruent, state the property that confirms that they are congruent.

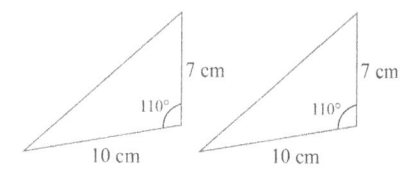

Applications

Solve.

11. An artist is attempting to stretch a canvas that is 3 by 4 feet. In order to check that the canvas is actually rectangular, the artist will measure the diagonal to determine if the corners form true right angles. What length should the diagonal of the stretched canvas be?

12. You and a friend are walking to class and want to figure out the height of the tree next to your building. Your friend is exactly 6 ft tall and casts 4 ft shadow. The tree casts a 12 ft shadow. How tall is the tree?

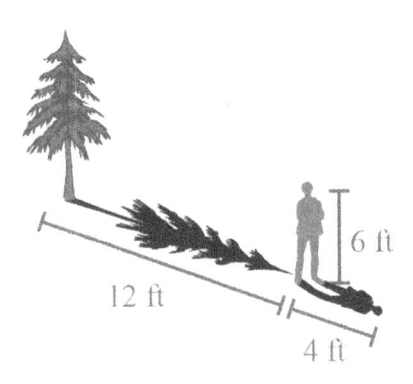

Writing & Thinking

13. Suppose you know the measurement of one of the acute angles of a right triangle and you know the length of the hypotenuse. Is this enough information to find the lengths of the other two sides? How would you go about finding this information?

6.7 Perimeter and Area

Polygon

A **polygon** is a closed plane figure, with _____ .

Each point where _____ .

Note: A **closed figure** begins and ends at the same point.

A **triangle** is _____ .

A **parallelogram** is _____ .

A **rectangle** is _____
_____ .

A **square** is _____ .

A **trapezoid** is _____ .

Perimeter

The **perimeter** P of a polygon is _____ .

Perimeter Formulas for Five Polygons

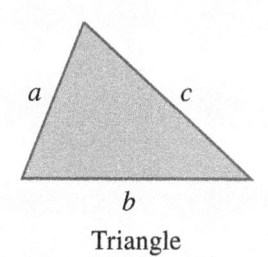

Triangle
$P =$

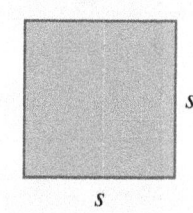

Square
$P =$

Rectangle
P =

Trapezoid
P =

Parallelogram
P =

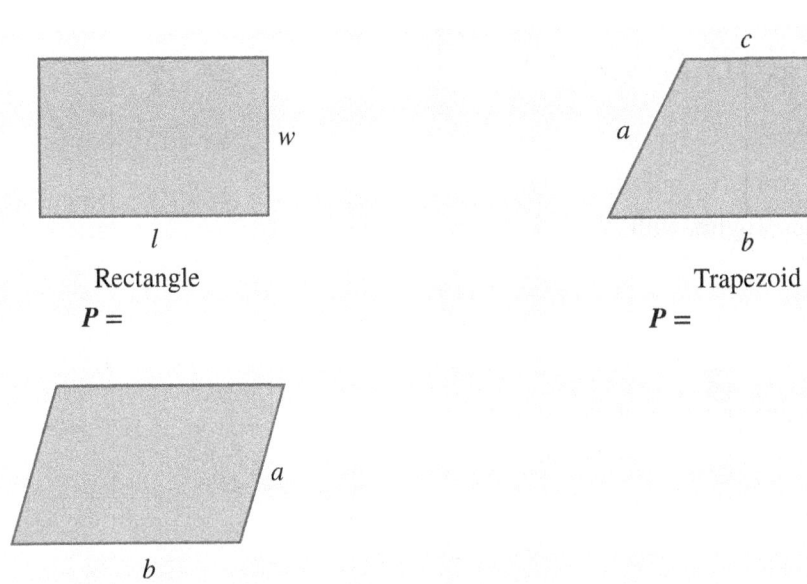

Note that each formula represents the sum of the lengths of the sides.

Circles

Circle: The set of all points in a plane that are

_____ .

Radius: The distance from the _____ .
(_____ is used to represent the radius of a circle.)

Diameter: The distance from _____
_____ . (_____ is used to represent the diameter of a circle
and _____ .)

Circumference: _____ .

The Circumference of a Circle

To find the circumference C of a circle, use one of the following formulas,

$\underline{\hspace{3cm}}$ and $\underline{\hspace{2cm}}$

where r is the $\underline{\hspace{1.5cm}}$ and d is the $\underline{\hspace{2cm}}$ of the circle.

Area is a measure of $\underline{\hspace{8cm}}$.

Area is measured in $\underline{\hspace{2.5cm}}$.

Area Formulas for Six Geometric Figures

Triangle

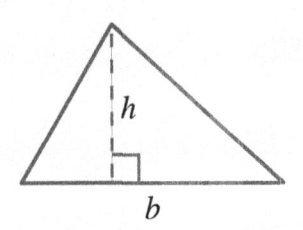

b (base), *h* (height)

$A = $ _____

Rectangle

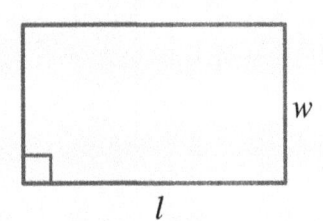

l (length), *w* (width)

$A = $ _____

Square

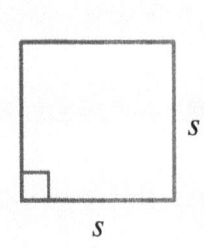

s, *s*

$A = $ _____

Parallelogram

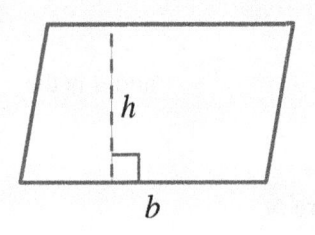

b, *h*

$A = $ _____

Trapezoid

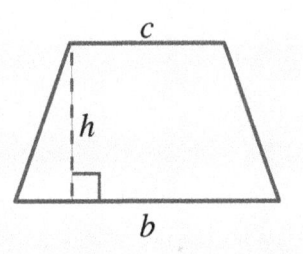

b, *c*, *h*

$A = $ _____

Circle

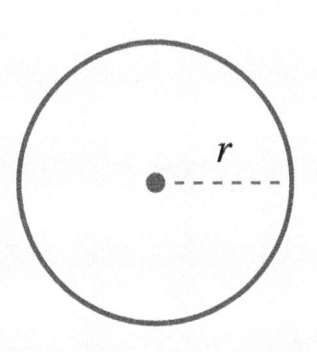

r

$A = $ _____

▶ Watch and Work

Watch the video for the example in the software and follow along in the space provided.

Example 9: Calculating the Area of a Trapezoid

Find the area of a trapezoid with altitude 6 in. and parallel sides of length 12 in. and 24 in.

Solution

✏ Now You Try It!

Use the space provided to work out the solution to the next example.

Example A: Calculating the Area of a Trapezoid

Find the area of a trapezoid with altitude 3 cm and parallel sides of length 9 cm and 15 cm.

Solution

6.7 Exercises

Concept Check

True/False. Determine whether each statement is true or false. If a statement is false, explain how it can be changed so the statement is true. (**Note:** There may be more than one acceptable change.)

1. **a.** Every square is a rectangle.

   b. Every rectangle is a square.

2. A trapezoid has only one pair of parallel lines.

3. The height of a triangle is the distance between the base and the vertex opposite the base.

4. The area formula for a triangle is $A = a + b + c$.

Practice

5. Find the perimeter of a square with sides of length 24 m.

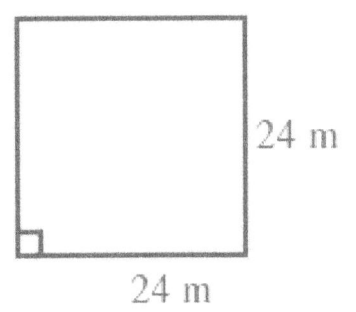

6. Find the perimeter of a triangle with sides of 20 cm, 12 cm, and 18 cm.

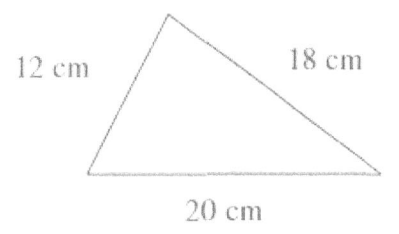

12 cm 18 cm
20 cm

7. Find the perimeter of a trapezoid with sides of 13 ft, 11 ft, 11 ft, and 14 ft.

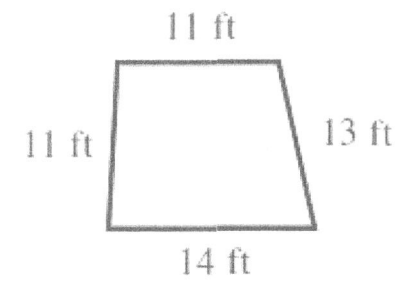

11 ft
11 ft   13 ft
14 ft

8. Find the circumference of a circle with a radius of 24 yd. Use $\pi = 3.14$. Round your answer to the nearest hundredth, if necessary.

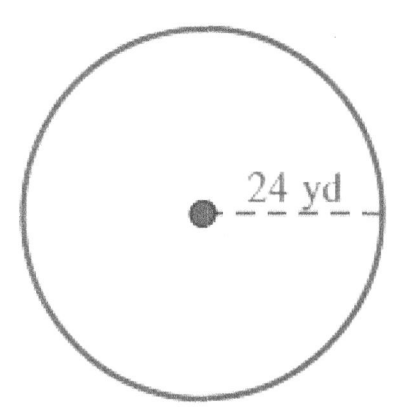

24 yd

256　　Chapter 6　　Geometric Pathways: Measurement & Geometry

9. Find the perimeter of the following figure with the indicated dimensions. Use $\pi = 3.14$. Round your answer to the nearest hundredth, if necessary.

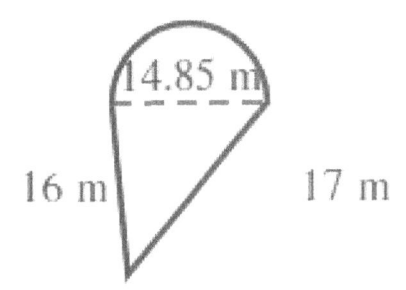

10. Find the area of a triangle with a base of 8 ft and height of 1 ft.

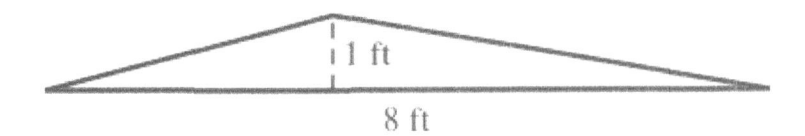

11. Find the area of a trapezoid with a base of 29 ft, a height of 5 ft, and top of 9 ft.

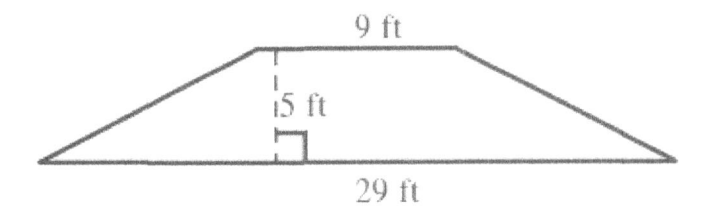

6.7   Perimeter and Area   257

12. Find the area of the following figure with the indicated dimensions. Round your answer to the nearest hundredth.

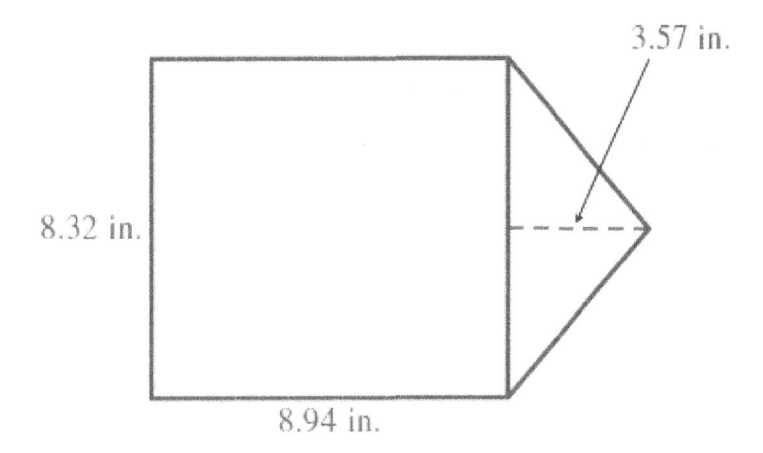

3.57 in.

8.32 in.

8.94 in.

13. Find the area of a circle with a radius of 18 m. Use $\pi = 3.14$. Round your answer to the nearest hundredth, if necessary.

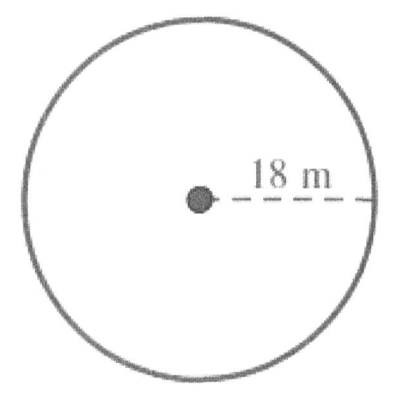

18 m

Applications

Solve.

14. A model sailboat has a triangular sail with the dimensions shown in the figure.

   a. What is the area of the sail?

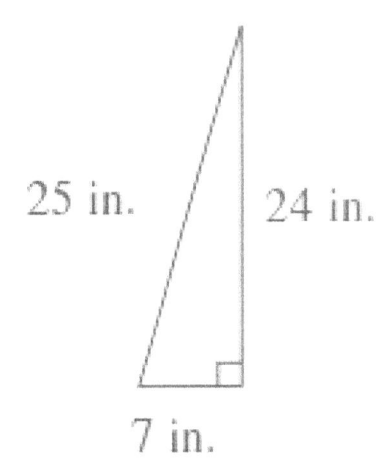

   b. What is the perimeter of the sail?

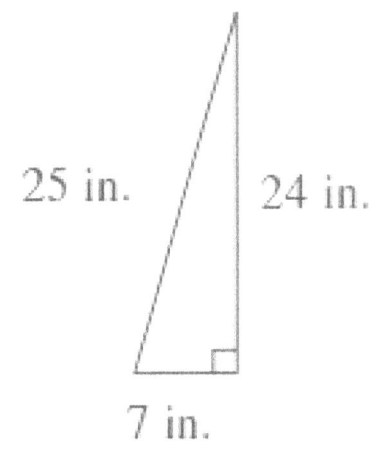

15. Cheryl is planting a five-sided lawn as shown in the figure. The lawn consists of a 20 foot by 10 foot rectangle and an attached 12 foot high triangle.

   a. What is the area of the lawn?

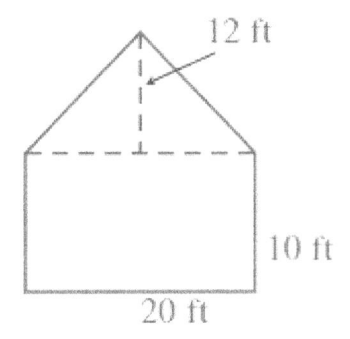

   b. If one pound of grass seed will cover 8 square feet, how many pounds will be necessary to cover the entire lawn?

Writing & Thinking

16. Name as many polygons as you can and include the number of sides for each one.

6.8 Volume and Surface Area

Volume is the measure of the _____. Volume is measured in _____ .

Volume Formulas for Five Geometric Solids

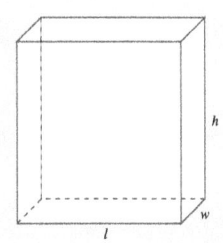

Rectangular solid

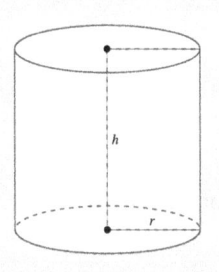

$V = $ _____

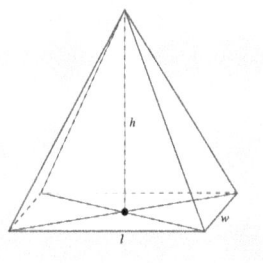

Rectangular pyramid

$V = $ _____

Right circular cylinder

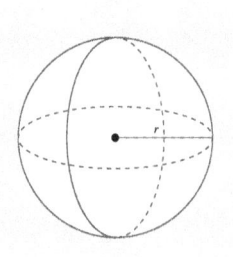

$V = $ _____

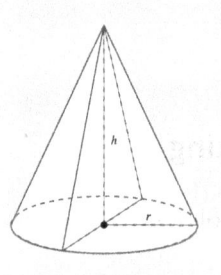

Right circular cone

$V = $ _____

Sphere

$V = $ _____

6.8 Volume and Surface Area

▶ Watch and Work

Watch the video for the example in the software and follow along in the space provided.

Example 1: Calculating the Volume of a Rectangular Solid

Calculate the volume of the rectangular solid with length 8 in., width 4 in., and height 1 ft.

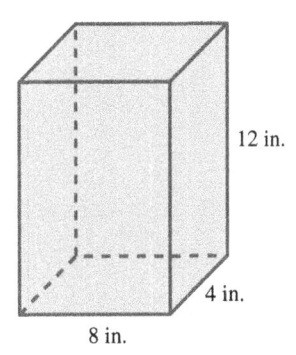

Solution

✏ Now You Try It!

Use the space provided to work out the solution to the next example.

Example A: Calculating the Volume of a Rectangular Solid

Calculate the volume of a rectangular solid with length 15 in., width 6 in., and height 9 in.

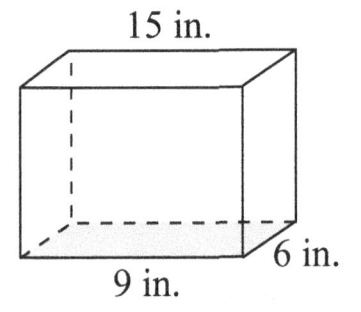

Solution

The **surface area** (*SA*) of a geometric solid is
_____ .

Surface Area Formulas for Three Geometric Solids

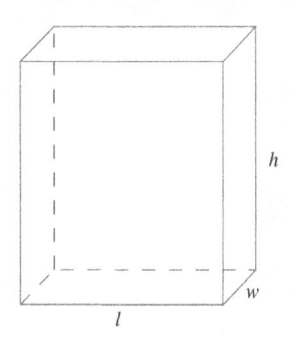

Rectangular solid

$SA = $ _____

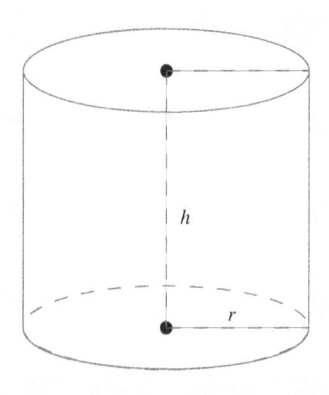

Right circular cylinder

$SA = $ _____

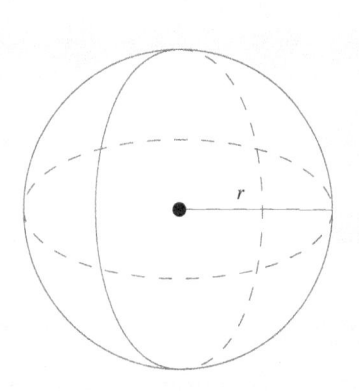

Sphere

$SA = $ _____

6.8 Exercises

Concept Check

True/False. Determine whether each statement is true or false. If a statement is false, explain how it can be changed so the statement is true. (**Note:** There may be more than one acceptable change.)

   1. To find the volume of a can of corn, the formula $V = \pi r^2 h$ would be used.

2. $V = lwh$ is the formula for the surface area of a rectangular solid.

3. The area of the paper label on a can of peaches is an example of surface area.

4. To find the volume of a rectangular solid, the areas of each surface are added together.

Practice

5. Find the volume of a rectangular solid with a length of 14 ft, a width of 8 ft, and a height of 24 ft. Round your answer to the nearest thousandth, if necessary.

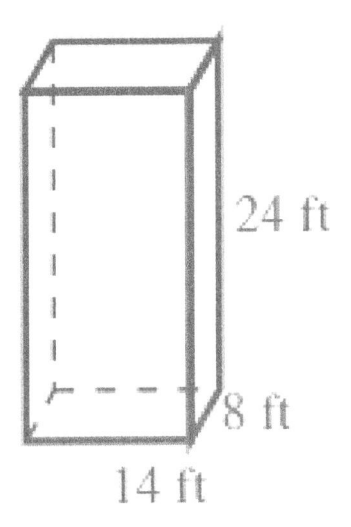

6. Find the volume of a sphere with a radius of 7 yd. (Note: Take the value of π as 3.14.) Round your answer to the nearest thousandth, if necessary.

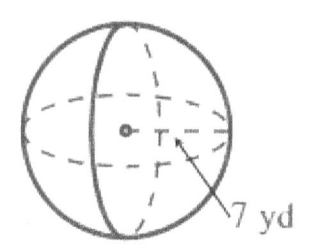

7. Find the volume of a right circular cone with a radius of 12 cm and a height of 14 cm. (Note: Take the value of π as 3.14.) Round your answer to the nearest thousandth, if necessary.

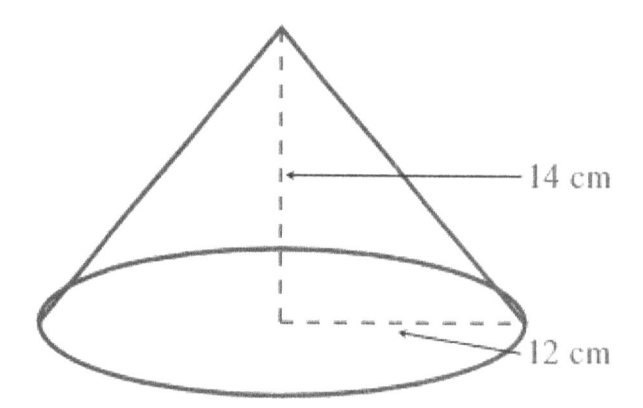

8. Find the volume of the following figure with the indicated dimensions. Round your answer to the nearest thousandth, if necessary.

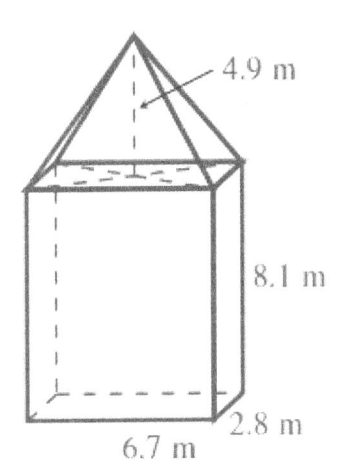

9. Calculate the surface area of the cube if the side length is 26 cm.

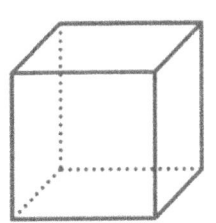

10. Calculate the surface area of the cylinder if the radius is 7 m and the height is 34 m. Round your answer to the nearest hundredth, if necessary. (Note: take the value of π as 3.14.)

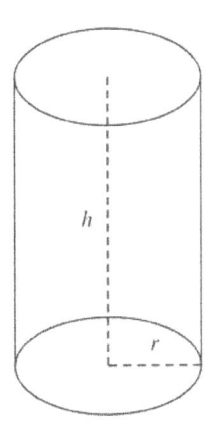

Applications

Solve.

11. A rectangular tent with straight sides has a pyramidal shaped roof. The dimensions of the rectangular portion are 12 ft long, 10 ft wide, and 6 ft high. The peak of the pyramid is 2 ft above the top edge of the walls. What is the volume of the inside of the tent?

12. Disposable paper drinking cups like those used at water coolers are often cone-shaped. Find the volume of such a cup that is 9 cm high with a 3.2 cm radius. Express the answer to the nearest milliliter.

Writing & Thinking

13. Discuss the type of units used for volume and explain why.

6.9 Right Triangle Trigonometry

Trigonometric Functions

If θ is an acute angle of a right triangle, then six trigonometric (or trig) functions are defined as follows.

_____   _____   _____

_____   _____   _____

where

hyp = _____

opp = _____

adj = _____

Common applications involving right triangles and trigonometry are of two basic types:

a. One acute angle and the length of one side are given and you are to find _____

_____ .

b. The lengths of two sides are given and you are to find _____ .

The process for solving these types of problems is known as _____ .

▶ Watch and Work

Watch the video for the example in the software and follow along in the space provided.

Example 3: Solving a Right Triangle

An architect is standing 100 feet from the base of a building and would like to know the height of the building. If he measures the angle of elevation (see figure) to be $69.5°$, what is the approximate height of the building (to the nearest tenth of a foot)?

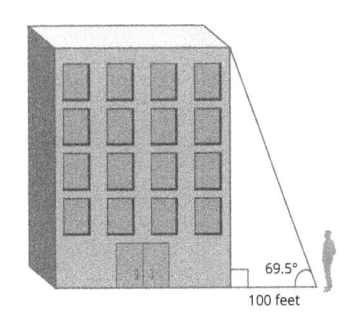

Solution

✏️ Now You Try It!

Use the space provided to work out the solution to the next example.

Example A: Solving a Right Triangle

An architect is standing 80 feet from the base of a building and would like to know the height of the building. If he measures the angle of elevation to be $72.5°$, what is the approximate height of the building (to the nearest tenth of a foot)?

Solution

6.9 Exercises

Concept Check

True/False. Determine whether each statement is true or false. If a statement is false, explain how it can be changed so the statement is true. (**Note:** There may be more than one acceptable change.)

1. The hypotenuse of a right triangle is the side opposite the right angle.

2. If θ is an acute angle of a right triangle, then $\sin \theta = \dfrac{\text{opp}}{\text{hyp}}$

3. If θ is an acute angle of a right triangle, then $\tan \theta = \dfrac{\text{adj}}{\text{hyp}}$

4. If the hypotenuse of a right triangle is 12 and the length of the side adjacent to angle θ is 5, then $\cos \theta = \dfrac{5}{12}$

Practice

5. The right triangle with hypotenuse of length 5 and legs of lengths 3 and 4 is shown with one of the acute angles labeled θ. Find the value of $\sin(\theta)$.

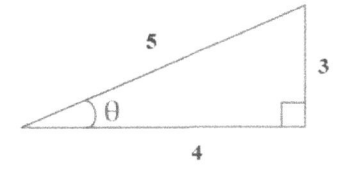

6. Use a calculator to find the value of the trig function. Round to the nearest hundredth.

$\tan(37°)$

7. Use a calculator to find the value of the trig function. Round to the nearest hundredth.

$\cos(60°)$

8. Determine the measure of the acute angle α in the given triangle. Round to the nearest tenth.

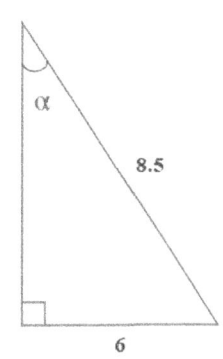

9. An architect is standing 200 feet from the base of a building and would like to know the height of the building. If he measures the angle of elevation (see figure) to be 60°, what is the approximate height of the building (to the nearest tenth of a foot)?

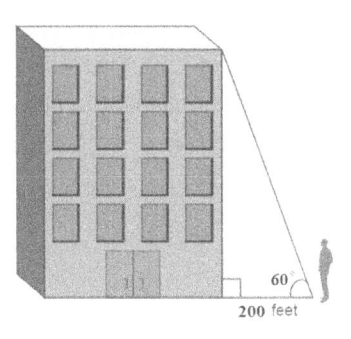

Writing & Thinking

10. Suppose you know the measurement of one of the acute angles of a right triangle and you know the length of the hypotenuse. Is this enough information to find the lengths of the other two sides? How would you go about finding this information?

Chapter 7
Pathways to Personal Finance

7.1 Percents

7.2 Simple and Compound Interest

7.3 Buying a Car

7.4 Buying a House

7.1 Percents

To change a decimal number to percent,

_____.

To change a percent to a decimal number:

_____.

To change a percent to a fraction (or mixed number):

_____.

To change a fraction (or mixed number) to a percent:

_____.

The Basic Formula $R \cdot B = A$

$R =$ _____

$B =$ _____

$A =$ _____

"of" means _____ .

"is" means _____ .

The relationship among R, B, and A is given in the equation

$$R \cdot B = A \quad (\text{or } A = R \cdot B).$$

▶ Watch and Work

Watch the video for the example in the software and follow along in the space provided.

Example 8: Application: Determining Commission

A saleswoman earns a salary of $1200 a month plus a commission of 8% on whatever she sells after she has sold $8000 in furniture. What did she earn the month she sold $25,000 worth of furniture?

Solution

Now You Try It!

Use the space provided to work out the solution to the next example.

Example A: Application: Determining Commission

Lynsay earns a salary of $1250 a month plus a commission of 5% on all electronics she sells at her job at the local computer store. What did she earn the month she sold $28,640 in electronics?

Solution

7.1 Exercises

Concept Check

True/False. Determine whether each statement is true or false. If a statement is false, explain how it can be changed so the statement is true. (**Note:** There may be more than one acceptable change.)

1. To change a decimal number to a percent, move the decimal point two places to the left and add the % sign.

2. When using the basic formula $R \cdot B = A$, the word "of" means to divide.

3. If an item is selling for a 35% discount, the customer will pay 65% of the original price.

4. A car was purchased in 1965 for $3800. It sold for $1200 in 2011. This is an example of depreciation.

Practice

5. Change the following decimal to a percent. Write your answer in percent form.

    0.012

6. Change the following percent to a decimal.

    11%

7. Change $1\frac{1}{4}\%$ to a fraction and reduce, if possible.

8. Change the following fraction to a percent. Write your answer in percent form. Round your answer to the nearest tenth of a percent, if necessary.

    $\frac{16}{25}$

9. Find the unknown quantity in the following percent problem using the formula $R \cdot B = A$. Round your answer to two decimal places, if necessary.

    111% of 189 is ____.

10. Find the unknown quantity in the following percent problem using the formula $R \cdot B = A$. Round your answer to two decimal places, if necessary.

    50% of ____ is 803.

11. Find the unknown quantity in the following percent problem using the formula $R \cdot B = A$. Round your answer to two decimal places, if necessary.

    ____% of 773 is 343.

Applications

Solve.

12. A sales clerk receives a monthly salary of $950 plus a commission of 7% on all sales over $3200. What did the clerk earn the month that he sold $13,500 in merchandise? Follow the problem-solving process and round your answer to the nearest cent, if necessary.

13. The population of white-tailed deer in a region was counted to be 321. The population in the same region the previous year was 300. Find the percent increase in the white-tailed deer population. Round your answer to the nearest hundredth, if necessary.

14. A few years ago, Sarah acquired a parcel of land valued at $13,800. Today, that same parcel of land has a value of $14,628. Find the percent increase in the property's value. Round your answer to the nearest hundredth, if necessary.

Writing & Thinking

15. A man weighed 200 pounds. He lost 20 pounds in 3 months. Then he gained back 20 pounds 2 months later.

    a. What percent of his weight did he lose in the first 3 months?

    b. What percent of his weight did he gain back?

    c. The loss and gain are the same, but the two percentages are different. Explain why.

7.2 Simple and Compound Interest

Simple Interest Formula

$$\text{Interest} = \text{Principal} \cdot \text{rate} \cdot \text{time},$$

Writing the formula using letters, we have $I = P \cdot r \cdot t$, where

$I =$ _____ (earned or paid),

$P =$ _____ (the amount invested or borrowed),

$r =$ _____ of interest (stated as an annual rate) in decimal number or fraction form, and

$t =$ _____ (years or fraction of a year).

▶ Watch and Work

Watch the video for the example in the software and follow along in the space provided.

Example 1: Application: Calculating Simple Interest

You want to borrow $2000 from your bank for one year. If the interest rate is 5.5%, how much interest would you pay?

Solution

✏ Now You Try It!

Use the space provided to work out the solution to the next example.

Example A: Application: Calculating Simple Interest

If you were to borrow $1500 at 8.5% for one year, how much interest would you pay?

Solution

278 Chapter 7 Pathways to Personal Finance

To Calculate Compound Interest

1. Use the formula _____ .

   Let $t = \dfrac{1}{n}$ where n _____ .

2. Add this interest to the _____ .

3. Repeat Steps _____
   _____ .

Compound Interest Formula

When interest is compounded, the total **amount** A accumulated (including principal and interest) is given by the formula

$$A = P\left(1 + \frac{r}{n}\right)^{nt}, \text{ where}$$

$P = $ _____ ,

$r = $ _____ ,

$t = $ _____

$n = $ _____ .

Total Interest Earned

To find the total interest earned on an investment that has earned interest by compounding, _____ .

$$I = \text{_____}$$

Inflation

The adjusted amount A due to **inflation** is

$$A = P(1+r)^t,$$

where

$P = $ _____ ,

$r = $ _____ ,

$t = $ _____ .

Depreciation

The **current value** of an item due to **depreciation** is

$$V = P(1-r)^t,$$

where

$P = $ _____ ,

$r = $ _____ ,

$t = $ _____ .

7.2 Exercises

Concept Check

True/False. Determine whether each statement is true or false. If a statement is false, explain how it can be changed so the statement is true. (**Note:** There may be more than one acceptable change.)

1. In the simple interest formula, the rate can be written as a decimal number or a fraction.

2. Simple interest can be compounded monthly or quarterly.

3. Interest cannot be earned on interest, only the principal.

4. Inflation can be treated in the same manner as simple interest.

Practice

5. What is the simple interest paid on $1750 at 4.5% for one year? Round your answer to the nearest cent.

6. How much interest would be paid on a loan of $3600 at 4% for 8 months? Round your answer to the nearest cent.

7. What principal would you need to invest at a rate of 6% to earn $250 in 10 months? Round your answer to the nearest cent.

Applications

Solve.

8. April wants to borrow $400 from her father and is willing to pay $16 in interest. Her father wants to charge an interest rate of 6%. How long can April keep the money?

9. If an account is compounded daily at 12%, how much interest will a principal of $30,300 earn in 53 months? Round your answer to the nearest cent.

10. Juan invested $12,800 at 13% to be compounded monthly. What will be the value of Juan's investment in 1 year? Round your answer to the nearest cent.

11. If the value of your house is $181,000 today and inflation is constant at 9.8% annually, what will be its value, to the nearest thousand dollars, in 10 years?

12. If a new car is valued at $24,400, what will its value be in 3 years if it depreciates 12.7% each year? Round your answer to the nearest cent, if necessary.

Writing & Thinking

13. Compare and contrast simple interest with compound interest.

7.3 Buying a Car

Terminology Used when Purchasing a Car

Purchase Price – _____
_____.

Principal – _____.

Down Payment –
_____.

Finance Charge – _____
_____.

Loan Repayment Formula

The formula to determine the monthly payment amount *PMT* for a fixed installment loan is

$$PMT = \underline{\hspace{4cm}},$$

where *P* is _____,

r is _____,

n is _____,

and *t* is _____.

▶ Watch and Work

Watch the video for the example in the software and follow along in the space provided.

Example 2: Calculating Car Payments and Finance Charges

Ada purchases a car that has a purchase price of $22,000 and makes a 25% down payment. She finances the remaining cost with a 4-year loan that has an annual percentage rate of 4.5%.

    a. Calculate the monthly payment for the car loan.

    b. Calculate the finance charge, or total interest paid, on the car loan after 4 years.

Solution

✏ Now You Try It!

Use the space provided to work out the solution to the next example.

Example A: Calculating Car Payments and Finance Charges

Kyle purchases a car that has a purchase price of $18,000 and makes a 20% down payment. He finances the remaining cost with a 4-year loan that has an annual percentage rate of 4%.

    a. Calculate the monthly payment for the car loan.

    b. Calculate the finance charge on the car loan after 4 years.

Solution

Finding the APR of a Loan

1. Calculate the _____ .

2. Find the finance charge per _____ .

   _____

3. Find the _____ that corresponds to the number of payments made and _____ _____ .

4. Find the percent listed _____ . That is the _____ .

Additional Expenses in Owning a Car

Auto Insurance – _____
_____ .

Operating Costs – Includes _____
_____ .

Repairs – Includes _____
_____ .

7.3 Exercises

Concept Check

True/False. Determine whether each statement is true or false. If a statement is false, explain how it can be changed so the statement is true. (**Note:** There may be more than one acceptable change.)

1. The purchase price of a car is the amount of interest paid over the life of the car loan.

2. Car loans are typically fixed installment loans with monthly payments.

3. The monthly payment for a car loan covers the interest earned on the loan and pays down a portion of the loan balance.

4. The APR of a loan takes into account any fees that are included in the amount financed.

Applications

Solve.

5. Amy bought a new car for $21,000. She paid a 10% down payment and financed the remaining balance for 36 months with an APR of 3.5%. Determine the monthly payment that Amy pays. Round your answer to the nearest cent, if necessary.

6. Jenna bought a new car for $24,000. She paid a 20% down payment and financed the remaining balance for 36 months with an APR of 3.5%. Determine the monthly payment that Jenna pays. Round your answer to the nearest cent, if necessary.

7. Jenna bought a new car for $32,000. She paid a 20% down payment and financed the remaining balance for 72 months with an APR of 6.5%. Assuming she makes monthly payments, determine the total interest Jenna pays over the life of the loan. Round your answer to the nearest cent, if necessary.

8. Jenna bought a new car for $35,000. She paid a 10% down payment and financed the remaining balance for 48 months with an APR of 4.5%. Assuming she makes monthly payments, determine the total interest Jenna pays over the life of the loan. Round your answer to the nearest cent, if necessary.

9. Max purchases a car and takes out a 4-year loan for $14,000. The loan officer tells him that with an interest rate of 6.5%, his monthly payment will be $345. What is the APR of the loan?

10. After purchasing a car, Eleanor adjusts her budget to cover most car-related expenses that may occur. The monthly payment for her loan is $297.64, the cost of car insurance for 6 months is $276.73, the estimated fuel cost per month is $145, and she wants to set aside $950 per year for repairs and other maintenance. How much should Eleanor budget per month for car expenses? Round your answer to the nearest cent.

Writing & Thinking

11. Create a list of operating costs and repairs that you might need to budget for when owning a car. Do some research to determine the approximate cost of each item on your list.

7.4 Buying a House

Terminology Used when Buying a House

Purchase Price: _____ .

Down Payment: _____ . Down payments are typically between _____ and _____ of the cost of the house, but can be as low as _____ in some situations.

Mortgage: _____ . These loans typically come from a bank or lending institution.

Private Mortgage Insurance (PMI): Required if the down payment is less than _____ . Can be paid upfront or _____ .

Points: _____ . One point is equal to _____ of the mortgage.

Escrow: _____ . These fees are included in your monthly payment.

Closing Costs: _____ _____ . This includes the down payment, lender fees, property tax, and homeowners insurance.

Loan Repayment Formula

The formula to determine the monthly payment amount PMT for a fixed installment loan is

$$PMT = \underline{\hspace{4cm}},$$

where P is _____ ,

r is _____ ,

n is _____ ,

and t is _____ .

▶ Watch and Work

Watch the video for the example in the software and follow along in the space provided.

Example 3: Determining Whether You Can Afford a House

You and your spouse make a combined $4200 per month and are looking to buy a house. You want to spend no more than 30% of your monthly budget on mortgage payments. You've saved $28,000 for a down payment and closing costs. You are interested in a house that is listed at $240,000. The estimated yearly property tax for the house is $3400 and the estimated homeowners insurance for a year is $1300.

a. Suppose you purchase the house for the price listed and make a 10% down payment. Determine the estimated monthly payment, including homeowners insurance and property tax, on a 30-year mortgage at 4.25% APR.

b. Does this monthly payment fit into your budget?

Solution

✏️ Now You Try It!

Use the space provided to work out the solution to the next example.

Example A: Determining Whether You Can Afford a House

You and your spouse make a combined $5000 per month and are looking to buy a house. You want to spend no more than 30% of your monthly budget on mortgage payments. You've saved $55,000 for a down payment and closing costs. You are interested in a house that is listed at $240,000. The estimated yearly property tax for the house is $2800 and the estimated homeowners insurance for a year is $1800.

a. Suppose you purchase the house for the price listed and make a 20% down payment. Determine the estimated monthly payment, including homeowners insurance and property tax, on a 30-year mortgage at 4% APR.

b. Does this monthly payment fit into your budget?

Solution

7.4 Exercises

Concept Check

True/False. Determine whether each statement is true or false. If a statement is false, explain how it can be changed so the statement is true. (**Note:** There may be more than one acceptable change.)

1. Mortgage insurance is required if you make a down payment of less than 20%.

2. An increase in the interest rate of a fixed installment loan can result in an increase in the monthly payment amount.

3. A general guideline is to spend at most 15% of your income (before taxes) on housing costs.

4. The first year of an amortization schedule illustrates that most of the monthly payment goes towards paying off the principal.

Practice

5. A house sells for $160,000 and the buyer makes a 15% down payment. Determine the down payment amount and the amount financed.

6. Calculate the monthly mortgage payment for a $155,000 mortgage that is financed for 30-years at 4.5% APR. Round to the nearest cent.

Applications

Solve.

7. Andrea is buying a house for $130,000. She plans to make a 20% down payment. Closing costs include $400 for 6 months of homeowners insurance, $900 for 6 months of property tax, $150 for the title fee, and $450 in transaction fees. Andrea also agreed to pay two points in exchange for a 0.5% reduction in interest rate.

   a. Determine the mortgage amount.

   b. Determine the amount of money Andrea needs to cover closing costs.

8. You and your spouse make $4800 per month and are looking to buy a house. You want to spend no more than 30% of your monthly budget on mortgage payments. You've saved $45,000 for a down payment and closing costs. You are interested in a house that is listed at $270,000. The estimated yearly property tax for the house is $2700 and the estimated homeowners insurance for a year is $1250.

   a. Suppose you purchase the house for the price listed and make a 15% down payment. Determine the estimated monthly payment, including homeowners insurance and property tax, on a 30-year mortgage at 4.5% APR.

   b. Does this monthly payment fit into your budget?

9. Complete the first three rows of the amortization table for a 30-year $200,000 mortgage that was financed at an APR of 4% and has a monthly payment of $954.83.

Payment Number	Interest Payment	Principal Payment	Mortgage Balance
1			
2			
3			

Writing & Thinking

10. Determine the total amount paid on a 30-year mortgage with principal $150,000 and a 4.5% APR. (Hint: There will be $30 \cdot 12 = 360$ monthly payments.) Would it be beneficial for the buyer to pay 2 points to lower the interest rate by 0.5%? If so, how much would the home owner save over the life of the loan?

Chapter 8
Pathways to Critical Thinking: Sets and Logic

8.1 Introduction to Sets

8.2 Venn Diagrams and Operations with Sets

8.3 Inductive and Deductive Reasoning

8.4 Logic Statements, Negations, and Quantified Statements

8.5 Compound Statements and Connectives

8.6 Truth Tables

8.1 Introduction to Sets

Sets

A **set** is a _____ . The items are called _____ _____ of the set.

There are three different ways to write sets. The first way is a _____ of the set. In this method, you simply give a _____ .

The second way to write a set is called _____ . In roster notation, you _____ _____ .

The third way to write a set is called _____ . It contains a _____ .

▶ Watch and Work

Watch the video for the example in the software and follow along in the space provided.

Example 4: Writing Sets

Here is a list of the world's five oceans: Arctic, Atlantic, Pacific, Indian, and Southern. Write set O, the set of the world's five oceans, using

   a. a word description,

   b. roster notation, and

   c. set-builder notation.

Solution

296 Chapter 8 Pathways to Critical Thinking: Sets and Logic

✏ Now You Try It!

Use the space provided to work out the solution to the next example.

Example A: Writing Sets

Here is a list of six basic types of quadrilaterals: parallelogram, rectangle, rhombus, square, kite, and trapezoid. Write set Q, the set of basic quadrilaterals, using

a. a word description

b. roster notation, and

c. set-builder notation.

Solution

The Symbols Element of and Not an Element of

$\in$ _____

$\notin$ _____

The Empty Set

The _____ is a set with no elements. It is represented by the symbol _____ or by writing _____ .

Subsets

If A and B are sets, B is a **subset** of A if _____ .

Proper Subsets

If _____, then B is a **proper subset** of A.

Note: If the subset B contains all the elements of A, then it is an _____.

Given a set that contains n number of elements, there are a total of _____ proper subsets.

8.1 Exercises

Concept Check

True/False. Determine whether each statement is true or false. If a statement is false, explain how it can be changed so the statement is true. (**Note:** There may be more than one acceptable change.)

1. Set-builder notation is the only way to write sets.

2. Given $A = \{1, 2, 3, 4, 5\}$, we can write that $5 \in A$ and $\{5\} \subseteq A$

3. Given $A = \{1, 2, 3, 4, 5\}$, A has 10 subsets.

4. The empty set can be represented by either ∅ or { }.

Practice

5. Write the set using the roster method.

    $A = \{x | x \text{ is an odd number and } 10 \leq x \leq 20\}$

6. Write the set using the roster method.

    Set A is the set of two-digit odd numbers less than 43 that are divisible by 3.

7. Let *A* equal the set of countries who have won fewer than 150 Gold medals. Write the set *A* using the roster method. Let the universal set consist of the 10 countries listed in the table. Use the country abbreviations US, SU, GB, G, F, I, S, C, R, and EG when writing the set.

Top 10 All-Time Olympic Medal Winning Countries				
Team	Gold	Silver	Bronze	Combined Total
United States (US)	1072	860	749	2681
Soviet Union (SU)	473	376	355	1204
Great Britain (GB)	246	276	284	806
Germany (G)	252	260	270	782
France (F)	233	254	293	780
Italy (I)	235	200	228	663
Sweden (S)	193	204	230	627
China (C)	213	166	147	526
Russia (R)	182	162	177	521
East Germany (EG)	192	165	162	519

Source: Wikipedia, s.v. "All-time Olympic Games medal table," accessed July 2014, http://en.wikipedia.org/wiki/All-time_Olympic_Games_medal_table

8. Fill in the blank with the symbol $\in$ or $\notin$ to make a true statement.

   $D = \{x | 0 < x < 10\}$

   a. 0 ___ D

   b. 5 ___ D

9. How many proper subsets of *A* are there if $A = \{6, 8, 14\}$? List them.

Writing & Thinking

10. Describe a situation where set builder notation would be more useful to write a set than roster notation.

8.2 Venn Diagrams and Operations with Sets

Universal Set

The **universal set**, U, contains _____

_____ .

Disjoint

Sets A and B are **disjoint** if _____ .

Intersection

The **intersection** of sets A and B is the set of all elements x such that _____

_____ . Mathematically, we write

_____ .

In other words, $A \cap B$ (read " _____ ") represents the set that contains the elements of set A _____ .

Union

The union of sets A and B is the set of all elements x such that _____

_____ . Mathematically, we write

_____ .

$A \cup B$ (read " _____ ") represents the set that contains elements that are only in set A, only in set B, or are in $A \cap B$.

Chapter 8 Pathways to Critical Thinking: Sets and Logic

▶ Watch and Work

Watch the video for the example in the software and follow along in the space provided.

Example 9: Finding the Union of Two Sets

$A = \{x | x \text{ is an even natural number less than or equal to } 10\}$

$B = \{x | x \text{ is a prime natural number less than or equal to } 10\}$

Find $A \cup B$.

Solution

✏️ Now You Try It!

Use the space provided to work out the solution to the next example.

Example A: Finding the Union of Two Sets

$A = \{x | x \text{ is an odd whole number less than or equal to } 10\}$

$B = \{x | x \text{ is a positive multiple of 3 less than or equal to } 10\}$

Find $A \cup B$.

Solution

Complement

A' (read " _____ ") represents the set that contains all the elements of the universal set _____ .

8.2 Exercises

Concept Check

True/False. Determine whether each statement is true or false. If a statement is false, explain how it can be changed so the statement is true. (**Note:** There may be more than one acceptable change.)

1. If the intersection of two sets is the empty set, the two sets are said to be disjoint.

2. The union of a set and its complement is the universal set.

3. A set and its complement always have at least one element in common.

4. The union of two sets contains only the elements that the two sets have in common.

Practice

5. List B using the roster method.

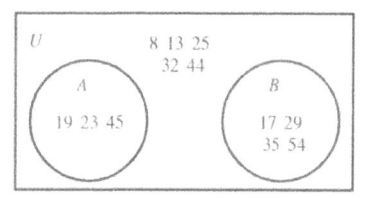

6. List U using the roster method.

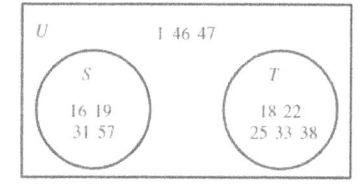

7. Use the given sets to find $A \cup B$.

   $A = \{1, 2, 3, 4, 5, 6, 7\}$

   $B = \{2, 4, 6, 8, 10, 12\}$

8. Use the given sets to find $A \cap B$.

   $A = \{1, 2, 3, 4, 5, 6, 7\}$

   $B = \{2, 4, 6, 8, 10, 12\}$

9. Let U consist of all beverages. Select the Venn diagram that represents the following two sets: sodas and teas.

   a. U — Sodas, Teas (two separate circles)

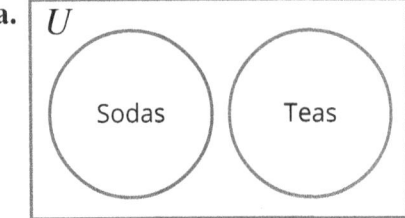

   b. U — Teas (large oval) containing Sodas

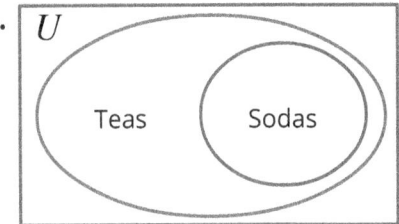

   c. U — Sodas, Teas (two overlapping circles)

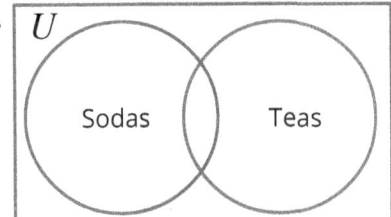

   d. U — Sodas (large oval) containing Teas

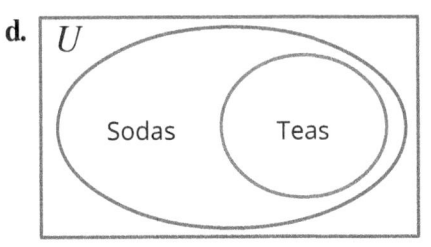

8.2 Venn Diagrams and Operations with Sets

10. Let U = { Pizza, Cheeseburger, Steak, Sandwich, Salad, Pasta, Soup }, $A = \{$Soup, Salad$\}$, and $B = \{$Soup, Pizza, Salad, Sandwich$\}$. Select the Venn diagram that represents U, A, and B.

a. U with A containing B inside

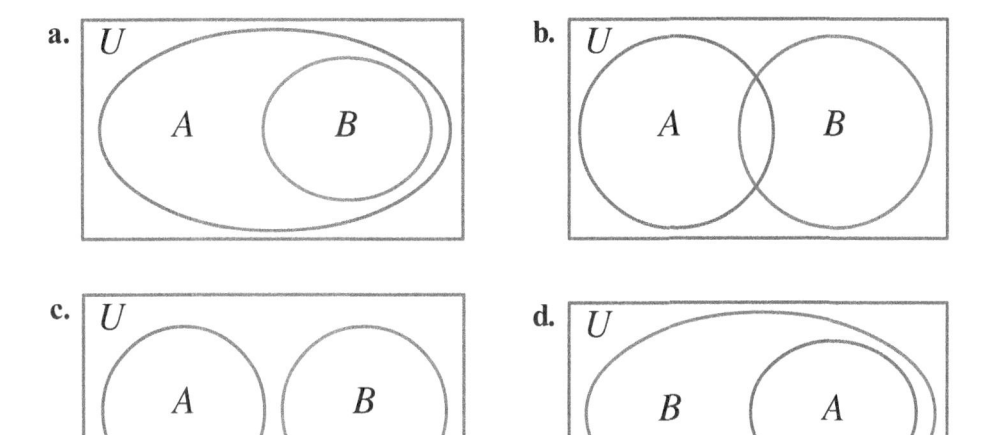

b. U with A and B overlapping

c. U with A and B separate

d. U with B containing A inside

11. If $A = \{a, u, t, o\}$ and U = { a, b, c, d, e, f, g, h, i, j, k, l, m, n, o, p, q, r, s, t, u, v, w, x, y, z }, find A'.

Applications

Solve.

12. In the following Venn diagram, U is the set of students in a class, A is the set of students who are majoring in psychology, and B is the set of students who are majoring in business. Determine which students are majoring in only psychology.

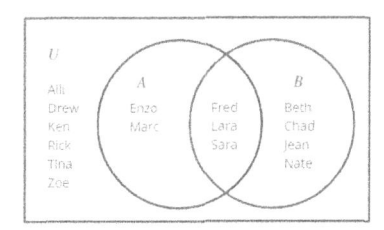

U: Alli, Drew, Ken, Rick, Tina, Zoe
A: Enzo, Marc
A ∩ B: Fred, Lara, Sara
B: Beth, Chad, Jean, Nate

13. In the following Venn diagram, U is the set of students in a class, A is the set of students who play soccer, and B is the set of students who play baseball. Determine which students play soccer and baseball.

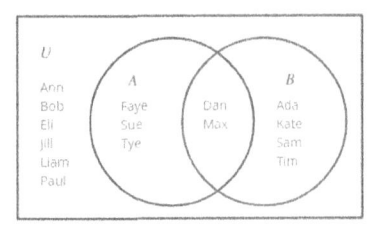

U: Ann, Bob, Eli, Jill, Liam, Paul
A: Faye, Sue, Tye
$A \cap B$: Dan, Max
B: Ada, Kate, Sam, Tim

8.3 Inductive and Deductive Reasoning

Inductive reasoning

Inductive reasoning is a process of _____

_____ .

Counterexample

A **counterexample** is an instance or time when _____ .

▶ Watch and Work

Watch the video for the example in the software and follow along in the space provided.

Example 3: Using Inductive Reasoning

Describe the pattern and use the pattern to draw the next figure in the sequence.

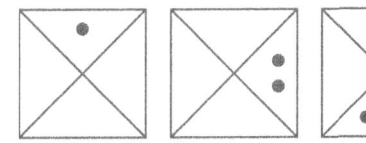

Solution

306 Chapter 8 Pathways to Critical Thinking: Sets and Logic

✏️ Now You Try It!

Use the space provided to work out the solution to the next example.

Example A: Using Inductive Reasoning

Describe the pattern and use the pattern to draw the next figure in the sequence.

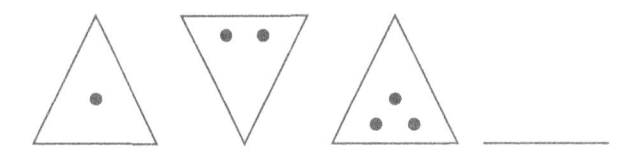

Solution

Deductive Reasoning

Deductive Reasoning is the process of _____
_____.

Inductive Reasoning versus Deductive Reasoning

Inductive reasoning uses _____
_____ , while
deductive reasoning uses _____ .

8.3 Exercises

Concept Check

True/False. Determine whether each statement is true or false. If a statement is false, explain how it can be changed so the statement is true. (**Note:** There may be more than one acceptable change.)

1. It is possible to reach the same conclusion using either inductive reasoning or deductive reasoning.

2. A conclusion reached using deductive reasoning is an educated guess.

3. Concluding that there will be an unannounced quiz in Biology at the end of each section because there have been unannounced quizzes in Biology at the end of the last three sections is an example of deductive reasoning.

4. Concluding that spaghetti cooks in water because all pasta cooks in water and spaghetti is a type of pasta is an example of deductive reasoning.

Practice

5. Identify a pattern in the list of numbers and use the pattern to find the next number.

    0,3,6,9,12,____

6. Identify a pattern in the list of numbers and use the pattern to find the next number.

    6,7,9,12,16,____

7. Determine if the statement is an example of inductive or deductive reasoning.

    All cookies in the box have pink frosting. You picked a cookie out of the box. Therefore, the cookie you picked has pink frosting.

8. Determine if the statement is an example of inductive or deductive reasoning.

Printer paper was on sale during the first week of January. Printer paper was also on sale during the first week of February. Therefore, printer paper will be on sale during the first week of March.

9. What conclusion can be drawn from these statements?

All trees require sunlight to survive. A silver birch is a type of tree.

   a. A silver birch will survive.

   b. A silver birch requires sunlight to survive.

   c. All trees are silver birches.

8.4 Logic Statements, Negations, and Quantified Statements

Logic

Logic is the process we use to validate a statement as _____.

Statements

A **statement** is a declaration that can be determined to be either _____, but not _____.

The **truth value** of a statement is the _____ that can be assigned to the statement.

Negation

A **negation** is the _____ e of a statement.

▶ Watch and Work

Watch the video for the example in the software and follow along in the space provided.

Example 4: Negating Statements

Determine the negations of the following statements.

p: I do not have time to finish my homework.

q: My homework is not due today.

Solution

Now You Try It!

Use the space provided to work out the solution to the next example.

Example A: Negating Statements

Determine the negations of the following statements.

p: I will not have a brownie for dessert.

q: My shoe is not tied.

Solution

Quantifiers and Quantified Statements

A **quantifier** indicates the _____ of the term it refers to.

A **quantified statement** is a statement whose subject is _____.

8.4 Exercises

Concept Check

True/False. Determine whether each statement is true or false. If a statement is false, explain how it can be changed so the statement is true. (**Note:** There may be more than one acceptable change.)

1. An opinion is not a valid statement.

2. All statements can be negated by adding the word "not".

3. "Not all trees are green" is the negation of the statement "All trees are green."

4. A statement can be both true and false at the same time.

Practice

5. Which of the following is an example of a mathematical statement?

   a. My boyfriend never sleeps because he sleeps all day.

   b. This milk has no chocolate in it.

   c. 23 is the coolest basketball jersey number.

   d. Nobody drives on Highway 16; there is too much traffic.

6. Which of the following is not an example of a mathematical statement?

   a. My girlfriend was born in Atlanta.

   b. I am driving behind a bus.

   c. 23 is the coolest basketball jersey number.

   d. Whole milk has 146 calories per cup.

7. Write the negation for the given statement.

   None of the dogs are barking.

8. Write the negation for the given statement.

   Yesterday was Saturday.

9. Write the negation of the following statement symbolically.

   p: Five is greater than three.

10. Determine which of the following are quantified statements. Select all that apply.

   a. Apples are a vegetable.

   b. Some of the pies are filled with apple.

   c. All pears are green.

   d. Pies are a type of pastry.

Writing & Thinking

11. Give an example to explain why the statement "Some of the fruit is ripe" is the negation of the statement "None of the fruit is ripe."

8.5 Compound Statements and Connectives

Types of Statements

A **simple statement** can be determined to be true or false, but _____.

A **compound statement** is created by connecting two or more simple statements with _____ _____.

A **logical connective** is a _____ (such as *and*, *or*, and *if/then*) that connects two or more simple statements.

Conjunctions

A **conjunction** is a compound statement made up of _____ combined with the logical connector _____.

Disjunctions

A **disjunction** is a compound statement made up of _____ combined with the logical connector _____.

Conditional Statements

A **conditional statement** (also called an _____) is a compound statement created when a simple statement *p* _____. These simple statements are joined by the *if/then* logical connector _____.

▶ Watch and Work

Watch the video for the example in the software and follow along in the space provided.

Example 4: Writing Conditional Statements

Use the given simple statements to write the indicated conditional statement.

p: I drive too fast.

q: I will get a ticket.

r: The shape is a rectangle.

s: The shape has four sides.

  a. p implies q

  b. r implies s

Solution

✏ Now You Try It!

Use the space provided to work out the solution to the next example.

Example A: Writing Conditional Statements

Use the given simple statements to write the indicated conditional statement.

p: I went to the store.

q: I bought a balloon.

r: The shape is a hexagon.

s: The shape has six sides.

  a. p implies q

  b. r implies s

Solution

Biconditional Statements

A **biconditional statement** is a compound statement created when a simple statement p implies _____ and _____.

These simple statements are joined by the *if and only if* logical connector. _____.

8.5 Exercises

Concept Check

True/False. Determine whether each statement is true or false. If a statement is false, explain how it can be changed so the statement is true. (**Note:** There may be more than one acceptable change.)

1. Simple statements contain logical connectives.

2. Conditional statements must be written using the if/then connective.

3. The symbol ⇔ means "if and only if".

4. Symbolic notation is a quick way to show the relationship between two simple statements.

Practice

5. Determine if the following statements are simple or compound.

   a. I can mow the lawn or I can take a nap.

   b. Avocados are a type of fruit.

c. The number 5 is an integer and is prime.

6. Determine whether the statement is a conjunction, a disjunction, a conditional statement, or a biconditional statement.

I will go to the store if and only if I need to buy milk.

7. Determine whether the statement is a conjunction, a disjunction, a conditional statement, or a biconditional statement.

I will read a book or I will watch a movie.

8. Consider the following statement: "If I press the button, then the alarm sounds."

Rewrite this statement using the alternative phrasing "p implies q."

9. Use the letters given to express the given compound statement in symbolic form.

Tom is outside or Tom is inside.

t: Tom is outside.

r: Tom is inside.

10. Use the letters given to express the given compound statement in symbolic form.

If the dog is not sleeping, then the doorbell rang.

t: The dog is sleeping.

r: The doorbell rang.

11. Use the given simple statements to write the compound statement $b \neg a$ in words.

a: It rained in April.

b: Flowers bloomed in May.

12. Use the given simple statements to write the compound statement $a \Rightarrow \sim b$ in words.

a: I like apples.

b: I like oranges.

Writing & Thinking

13. Rewrite the following paragraph as a series of symbolic statements. Be sure to define the simple statements.

If it is raining, then the dog is muddy. The dog will get a bath if and only if the dog is muddy. The dog will get a bath and not go back outside.

8.6 Truth Tables

Truth Table

A **truth table** describes the possible truth values of a _____. This is accomplished by organizing all the possible combinations of truth values of the _____ _____.

▶ Watch and Work

Watch the video for the example in the software and follow along in the space provided.

Example 1: Constructing a Truth Table for a Negation

Construct a truth table for $\sim p$.

Solution

8.6 Truth Tables

✏️ Now You Try It!

Use the space provided to work out the solution to the next example.

Example A: Constructing a Truth Table for the Negation

Construct a truth table for the negation of the following statement.

p: The soup is hot

Solution

Truth Values of Conjunctions and Disjunctions

The conjunction $p \wedge q$ is _____ only when both p and q are _____. Otherwise, the conjunction is _____.

The disjunction $p \vee q$ is _____ only when both p and q are _____. Otherwise, the disjunction is _____.

Truth Value of Conditional Statements

The conditional statement $p \Rightarrow q$ is _____ only when p is true and q is _____. Otherwise, the conditional statement is _____.

8.6 Exercises

Concept Check

True/False. Determine whether each statement is true or false. If a statement is false, explain how it can be changed so the statement is true. (**Note:** There may be more than one acceptable change.)

1. A compound statement can have different truth values depending on the truth values of the simple statements it contains.

2. A disjunction $p \vee q$ is always true regardless of the truth values of p and q.

3. The number of columns in a truth table is increased when negations are involved.

4. A truth table must contain every possible combination of true and false for the simple statements involved.

Practice

5. Create a truth table for the following expression.

$w \wedge z$

Truth Table		
w	z	$w \wedge z$

6. Create a truth table for the following expression.

$m \wedge {\sim}n$

Truth Table			
m	n	${\sim}n$	$m \wedge {\sim}n$

7. Create a truth table for the following expression.

$(a \wedge b) \Rightarrow c$

Truth Table				
a	b	c	$a \wedge b$	$(a \wedge b) \Rightarrow c$

8. Use variables to rewrite the given compound statement.

If it is sunny outside, then Amy is going to the beach.

9. Use variables to rewrite the given compound statement.

If Tom is on the beach and isn't wearing sunscreen, then he will get a sunburn.

Chapter 9
Statistical Pathways: Introduction to Probability

9.1 Introduction to Probability

9.2 The Addition Rules of Probability and Odds

9.3 The Multiplication Rules of Probability and Conditional Probability

9.4 The Fundamental Counting Principle and Permutations

9.5 Combinations

9.6 Using Counting Methods to Find Probability

9.1 Introduction to Probability

Terms Related to Probability

A **trial** is a _____ .

An **outcome** of a trial is the _____
_____ .

The **sample space** is the _____
_____ .

An **event** is a _____ .

Random Experiment

A **random experiment** is defined as any activity or phenomenon that meets the following conditions.

1. There is _____ for each trial of the experiment.

2. The outcome of the experiment is _____ .

3. The set of all distinct outcomes of the experiment
_____ , denoted by S.

Empirical Probability

The **empirical probability** of an event (E) is obtained by performing a random experiment and computing the _____
_____ .

$P(E) = $ _____ .

▶ Watch and Work

Watch the video for the example in the software and follow along in the space provided.

Example 5: Calculating Empirical Probability

A basketball player has made 300 of 420 free throw attempts.

a. What is the empirical probability that the player will make a free throw attempt?

b. What is the probability he will not make a free throw attempt?

Solution

✏️ Now You Try It!

Use the space provided to work out the solution to the next example.

Example A: Calculating Empirical Probability

A basketball player has made 150 of 215 free throw attempts.

a. What is the empirical probability that the player will make a free throw attempt?

b. What is the probability he will not make a free throw attempt?

Solution

Classical Probability

Using the **classical approach** to probability, the probability of an event E, denoted $P(E)$, is given by

$P(E) = $ _____ .

When using the classical approach, always make sure that _____

_____ .

9.1 Exercises

Concept Check

True/False. Determine whether each statement is true or false. If a statement is false, explain how it can be changed so the statement is true. (**Note:** There may be more than one acceptable change.)

1. In an experiment where a coin is tossed several times, the trials are a head or a tail.

2. In an experiment where a standard die is rolled, the sample space would be $S = \{1, 2, 3, 4, 5, 6\}$.

3. An empirical probability is a probability that is determined by observing a random experiment.

4. Classical probability can be measured as a simple proportion: the number of trials in an experiment divided by the number of outcomes in the sample space.

Practice

5. Write out the sample space for the given experiment. Use the letter H to indicate heads and T for tails.

   3 coins are tossed.

6. Write out the sample space for the given experiment. Use the following letters to indicate each choice: M for mushrooms, A for asparagus, S for shrimp, B for bacon, F for French, and V for vinaigrette.

   When deciding what you want to put on a salad for dinner at a restaurant, you will choose one of the following extra toppings: mushrooms, asparagus. Also, you will add one of the following meats: shrimp, bacon. Lastly, you will decide on one of the following dressings: French, vinaigrette.

Applications

Solve.

7. A mail order company classifies its customers by gender and location of residence. The research department has gathered data from a random sample of 1878 customers. The data is summarized in the table below.

Gender and Residence of Customers		
Residence	**Males**	**Females**
Apartment	283	221
Dorm	114	220
With Parent(s)	94	195
Sorority/Fraternity House	146	249
Other	69	287

   What is the probability that a customer lives with their parents? Express your answer as a fraction or a decimal number rounded to four decimal places.

8. A credit card company classifies its customers by gender and location of residence. The research department has gathered data from a random sample of 1490 customers. The data is summarized in the table below.

Gender and Residence of Customers		
Residence	**Males**	**Females**
Apartment	192	222
Dorm	86	59
With Parent(s)	174	117
Sorority/Fraternity House	275	58
Other	57	250

What is the probability that a customer is female and lives in a dorm? Express your answer as a fraction or a decimal number rounded to four decimal places.

9. A standard die is rolled. Find the probability that the number rolled is less than 2. Express your answer as a fraction in lowest terms or a decimal rounded to the nearest millionth.

10. Carter has 123 songs on a playlist. He's categorized them in the following manner: 11 gospel, 25 country, 21 blues, 6 folk, 16 pop, 14 jazz, and 30 rock. If Carter begins listening to his playlist on shuffle, what is the probability that the first song played is a gospel song? Express your answer as a fraction in lowest terms or a decimal rounded to the nearest millionth.

Writing & Thinking

11. Suppose that in trying to find the probability of a coin landing on heads three times in a row, you performed 100 trials of tossing a coin three times in while your friend performed the same trial 500 times. Which of you would be more likely to obtain an empirical probability closest to the actual probability?

9.2 The Addition Rules of Probability and Odds

Two events are mutually exclusive if they _____.

> ### Addition Rule for Probability of Mutually Exclusive Events
>
> If two events A and B are mutually exclusive,
>
> $$P(A \text{ or } B) = P(A) + P(B).$$
>
> In set notation, this can be written _____.
>
> In words, if two events are mutually exclusive, you can simply
>
> _____.

> ### Addition Rule for Probability of Non-mutually Exclusive Events
>
> If A and B are not mutually exclusive events, then
>
> $$P(A \text{ or } B) = P(A) + P(B) - P(A \text{ and } B),$$
>
> where $P(A \text{ and } B)$ represents the probability of the intersection of events A and B.
>
> In set notation, this can be written _____.

▶ Watch and Work

Watch the video for the example in the software and follow along in the space provided.

Example 9: Finding the Probability of Non-mutually Exclusive Events

You have a container with balls numbered between 1 and 20. If you randomly draw a ball from the container, what is the probability that the number on the ball will be divisible by three or divisible by four?

Solution

9.2   The Addition Rules of Probability and Odds   331

✏️ Now You Try It!

Use the space provided to work out the solution to the next example.

Example A: Finding the Probability of Non-mutually Exclusive Events

You have a container with balls numbered between 1 and 15. If you randomly draw a ball from the container, what is the probability that the number on the ball will be divisible by two or divisible by three?

Solution

Probability Law for Complementary Events

Let A be some event and A' be the complement of A.

Then _____ .

The sum of the probabilities of an event and its complement must equal _____ .

Odds

The **odds in favor** of an event A occurring are given by _____ .

The odds against an event A occurring are given by _____ .

Finding Probability Given the Odds

If we are given the odds in favor of an event A as n to m, then the probability of event A can be calculated by

$$P(A) = \underline{\qquad\qquad}.$$

9.2 Exercises

Concept Check

True/False. Determine whether each statement is true or false. If a statement is false, explain how it can be changed so the statement is true. (**Note:** There may be more than one acceptable change.)

1. If it is possible for two events to occur at the same time, then those two events are mutually exclusive.

2. If a certain team is given 1 to 3 odds of winning a game, this means that for every dollar bet on the team, a person would win $3 if the team wins.

3. An event A and its complement A' make up the entire sample space.

4. If you have 1 to 5 odds in favor of winning a game, then the probability of winning is $\frac{1}{5}$.

Practice

5. Determine whether the following events are mutually exclusive.

Choosing a red card or a black card out of a standard deck of cards.

6. Determine whether the following events are mutually exclusive.

Choosing a student who is a junior or a chemistry major from a nearby university to participate in a research study.

Applications

Solve.

7. Two dice are rolled. What is the probability that the sum of the numbers rolled is either 2 or 3? Express your answer as a fraction in lowest terms or a decimal rounded to the nearest millionth.

8. A bag of 11 marbles contains 6 marbles with red on them, 3 with blue on them, 6 with green on them, and 4 with red and green on them. What is the probability that a randomly chosen marble has either green or red on it? Note that these events are not mutually exclusive. Express your answer as a fraction in lowest terms or a decimal rounded to the nearest millionth.

9. Suppose the probability of school being canceled for snow is 0.75. What is the probability that school will not be canceled?

10. Suppose the probability of a candidate being elected to student government is 0.10. What are the odds of being elected? Express your answer in the form $a : b$.

11. If your baseball team has 4 : 1 odds in favor of winning this weekend's game, what is the probability implied by these odds?

Writing & Thinking

12. Suppose we conducted an experiment in which we rolled a two dice and were interested in the event A: rolling a number divisible by 3 and B: rolling a number greater than 7. Describe the error made if we found the probability of rolling a number divisible by 3 or greater than 7 to be $\frac{9}{12} = \frac{3}{4}$.

9.3 The Multiplication Rules of Probability and Conditional Probability

Conditional Probability

The probability that an event will occur _____, is a **conditional probability**.

Rule for Conditional Probability

The conditional probability of event A occurring given that event B has already occurred is

_____

or, using set notation,

_____

given that $P(B) \neq 0$.

The notation $P(A|B)$ is read as "_____." The vertical bar within a probability statement will always mean *given*.

Independent

Two events, A and B, are **independent** if and only if

_____.

Multiplication Rule for Independent Events

If two events, A and B, are **independent**, then

_____.

If n events, $A_1, A_2, \ldots, A_n$, are independent, then

_____.

▶ Watch and Work

Watch the video for the example in the software and follow along in the space provided.

Example 6: Using the Multiplication Rule for Independent Events

Choose two cards from a standard deck of 52 cards. Assume that before choosing the second card, you will replace the first card and shuffle. What is the probability of choosing a king and then a queen?

Solution

✏ Now You Try It!

Use the space provided to work out the solution to the next example.

Example A: Using the Multiplication Rule for Independent Events

Choose two cards from a standard deck of 52 cards. Assume that before choosing the second card, you will replace the first card and shuffle. What is the probability of choosing an ace and then a diamond?

Solution

Multiplication Rule for Dependent Events

If two events, A and B, are dependent, then

_____ .

9.3 Exercises

Concept Check

True/False. Determine whether each statement is true or false. If a statement is false, explain how it can be changed so the statement is true. (**Note:** There may be more than one acceptable change.)

1. The probability of an event B, given that event A has occurred, is always equal to the probability of an event A, given that event B has occurred.

2. Two events are said to be independent if knowing that one event occurred does not provide information regarding whether the other event has occurred.

3. The probability that mutually dependent events occur at the same time is the product of their probabilities.

4. Flipping a coin and rolling a die are independent events

Applications

Solve.

5. Two cards are drawn without replacement from a standard deck of 52 playing cards. What is the probability of choosing a queen for the second card drawn, if the first card, drawn without replacement, was a king? Express your answer as a fraction or a decimal number rounded to four decimal places.

6. A box contains 13 green marbles and 15 white marbles. If the first marble chosen was a green marble, what is the probability of choosing, without replacement, a white marble? Express your answer as a fraction or a decimal number rounded to four decimal places.

7. Two cards are drawn without replacement from a standard deck of 52 playing cards.

   What is the probability of choosing a club and then, without replacement, a red card? Express your answer as a fraction or a decimal number rounded to four decimal places.

8. Suppose you like to keep a jar of change on your desk. Currently, the jar contains the following:

   18 Pennies   18 Dimes

   16 Nickels   22 Quarters

   What is the probability that you reach into the jar and randomly grab a nickel and then, without replacement, a dime? Express your answer as a fraction or a decimal number rounded to four decimal places.

Writing and Thinking

9. Explain in your own words why $P(A \text{ and } B)$ is only equal $P(A) \cdot P(B)$ to if A and B are independent.

9.4 The Fundamental Counting Principle and Permutations

The Fundamental Counting Principle

For two events, if there are m possible outcomes for the first event and n possible outcomes for the second, then there are _____ ways for the two events to occur in the given order.

Similarly, for three events, if there are m possible outcomes for the first event, n possible outcomes for the second event, and p possible outcomes for the third event, then there are _____ ways for the three events to occur in the given order.

In general, the total number of ways several events can occur in a given order is found by _____

_____ .

Permutation

A **permutation** is an _____ of the elements of a set where the order matters.

n Factorial ($n!$)

For any positive integer n, the factorial of n, denoted as $n!$, is the _____

_____ .

$$n! = n(n-1)(n-2)\ldots(3)(2)(1)$$

$n!$ is read as "n factorial."

Note: The factorial of zero is a special case. It is defined as _____ .

Number of Permutations of n Elements

The number of permutations of n elements can be calculated by the following formula.

_____

_____

That is, n elements can be arranged in $n!$ ways.

Number of Permutations of *n* Elements Taken *r* at a Time

The symbol $_nP_r$ denotes the number of permutations of *n* elements taken *r* at a time.

$$_nP_r = \underline{\hspace{5in}}$$

Note: Other notations for permutations are P_r^n and $P(n, r)$.

▶ Watch and Work

Watch the video for the example in the software and follow along in the space provided.

Example 4: Calculating the Number of Permutations of *n* Elements Taken *r* at a Time

A sailor has 7 different flags he can signal with. How many signals can he send using 3 flags at a time?

Solution

✏ Now You Try It!

Use the space provided to work out the solution to the next example.

Example A: Calculating the Number of Permutations of *n* Elements Taken *r* at a Time

A sailor has 8 different flags she can signal with. How many signals can he send using 3 flags at a time?

Solution

9.4 Exercises

Concept Check

True/False. Determine whether each statement is true or false. If a statement is false, explain how it can be changed so the statement is true. (**Note:** There may be more than one acceptable change.)

1. For two events, if there are m possible outcomes for the first event and n possible outcomes for the second event, then there are $m + n$ ways for the two events to occur in the given order.

2. The product of the positive integers from 5 to 1 can be represented by 5!.

3. Each ordering of a set of elements is call a factorial.

4. 0! is undefined.

Practice

5. Evaluate the following expression.

   $$\frac{8!}{3!}$$

6. Evaluate the following expression.

   $$\frac{13!}{4!\,(13-4)!}$$

Applications

Solve.

7. A value meal package at Ron's Subs consists of a drink, a sandwich, and a bag of chips. There are 4 types of drinks to choose from, 5 types of sandwiches, and 3 types of chips. How many different value meal packages are possible?

8. User passwords for a certain computer network consist of 2 letters followed by 2 numbers. How many different passwords are possible? Repetition is allowed.

9. A doctor visits her patients during morning rounds. In how many ways can the doctor visit 4 patients during the morning rounds?

10. A mother duck lines her 5 ducklings up behind her. In how many ways can the ducklings line up?

11. A coordinator will select 4 songs from a list of 5 songs to compose an event's musical entertainment lineup. How many different lineups are possible?

12. The newly elected president needs to decide the remaining 3 spots available in the cabinet he/she is appointing. If there are 12 eligible candidates for these positions (where rank matters), how many different ways can the members of the cabinet be appointed?

Writing & Thinking

13. To determine how many four-digit numbers can be formed using digits 1, 2, 3, 4, 5, and 6, we can find the number of permutations of 6 elements taken 4 at a time: $\frac{6!}{(6-4)!} = \frac{6!}{2!} = 6 \cdot 5 \cdot 4 \cdot 3$. How does this relate to using the fundamental principle of counting?

9.5 Combinations

> **Combination**
>
> A **combination** is a collection of some (or all) of the elements of a set _____.

> **Number of Combinations of *n* Elements Taken *r* at a Time**
>
> The symbol $_nC_r$ denotes the number of combinations of *n* elements taken *r* at a time.
>
> $$_nC_r = \phantom{\rule{3cm}{0.4pt}}$$

▶ Watch and Work

Watch the video for the example in the software and follow along in the space provided.

Example 4: Calculating Combinations

A restaurant runs a deal where you can get a large 3-topping pizza for $14. If there are 16 different toppings, how many different 3-topping pizzas can you choose from?

Solution

✏️ Now You Try It!

Use the space provided to work out the solution to the next example.

Example A: Calculating Combinations

A restaurant runs a deal where you can get a large 4-topping pizza for $16. If there are 12 different toppings, how many different 4-topping pizzas can you choose from?

Solution

9.5 Exercises

Concept Check

True/False. Determine whether each statement is true or false. If a statement is false, explain how it can be changed so the statement is true. (**Note:** There may be more than one acceptable change.)

1. For permutations, the order of the elements is important. In a combination, the order doesn't matter.

2. The number of combinations of n elements taken r at a time is greater than the number of permutations of n elements taken r at a time.

3. The total number of combinations of n elements taken r at a time is symbolized $_rC_n$.

4. Determining how many ways a hand of cards can be dealt is a combination problem.

Practice

5. Determine whether each situation is a combination or a permutation. Explain why.

   a. A company is sponsoring 6 employees for a 5K race. Twelve employees are interested in running the race. How many different outcomes are there for the company to pick 6 employees to sponsor?

b. Trophies are awarded to the first 5 people who finish the 5k race. There are trophies for 1st place, 2nd place, 3rd place, 4th place, and 5th place. If 100 people are participating in the race, how many possible outcomes are there for winners?

Applications

Solve.

6. How many ways can Aileen choose 3 pizza toppings from a menu of 7 toppings?

7. A person tosses a coin 6 times. In how many ways can he get 3 tails?

8. Ben is studying photography, and he was asked to submit 5 photographs from his collection to exhibit at the fair. He has 13 photographs that he thinks are show worthy. In how many ways can the photographs be chosen?

Writing & Thinking

9. Since each combination of r elements has $r!$ permutations, the product $r! \cdot {}_nC_r$ represents the number of permutations of n elements taken r at a time. In other words, $r! \cdot {}_nC_r = {}_nP_r$. Show how this can be used to find that ${}_nC_r = \dfrac{n!}{r!\,(n-r)!}$.

346 Chapter 9 Statistical Pathways: Introduction to Probability

9.6 Using Counting Methods to Find Probability

Review of Probability Terms

A **trial** is a _____.

An **outcome** of a trial is the _____

_____.

The **sample space** is the _____

_____.

An **event** (E) is a _____.

The **probability of an event** E occurring is defined as

$$P(E) = \underline{\hspace{10cm}}.$$

▶ Watch and Work

Watch the video for the example in the software and follow along in the space provided.

Example 2: Calculating Probabilities Using Combinations

A math club is sending 3 random members to a conference. There are 9 math majors and 5 statistics majors in the club. What is the probability that only math majors are selected to go to the conference?

Solution

9.6 Using Counting Methods to Find Probability

✏️ Now You Try It!

Use the space provided to work out the solution to the next example.

Example A: Calculating Probabilities Using Combinations

A science club is sending 4 random members to a conference. There are 10 chemistry majors and 6 physics majors in the club. What is the probability that only chemistry majors are selected to go to the conference?

Solution

9.6 Exercises

Applications

Solve.

1. Find the probability of obtaining exactly one head when flipping three coins. Express your answer as a fraction in lowest terms or a decimal rounded to the nearest millionth.

2. A local pizza parlor has the following list of toppings available for selection. The parlor is running a special to encourage patrons to try new combinations of toppings. They list all possible two-topping pizzas (2 distinct toppings) on individual cards and give away a free pizza every hour to a lucky winner. Find the probability that the first winner randomly selects the card for the pizza topped with chicken fingers and mushrooms. Express your answer as a fraction in lowest terms or a decimal rounded to the nearest millionth.

   Pizza Toppings: Green Peppers, Onions, Kalamata Olives, Sausage, Mushrooms, Black Olives, Pepperoni, Spicy Italian Sausage, Roma Tomatoes, Green Olives, Ham, Grilled Chicken, Jalapeño Peppers, Banana Peppers, Beef, Chicken Fingers

3. It is presentation day in class and your instructor is drawing names from a hat to determine the order of the presentations. If there are 14 students in the class, what is the probability that the first 2 presentations will be by Harmony and Ben, in that order? Express your answer as a fraction in lowest terms or a decimal rounded to the nearest millionth.

4. A committee of 3 is being formed randomly from the employees at a school: 4 administrators, 30 teachers, and 4 staff. What is the probability that all 3 members are teachers? Express your answer as a fraction in lowest terms or a decimal rounded to the nearest millionth.

Writing & Thinking

5. Describe a scenario that would require the use of permutations to determine the size of the sample space or the number of outcomes of an event E. Describe a scenario that would require the use of combinations to determine the size of the sample space or the number of outcomes of an event E.

Chapter 10
Statistical Pathways: Introduction to Statistics

10.1 Collecting Data

10.2 Organizing and Displaying Data

10.3 Measures of Center

10.4 Measures of Dispersion and Percentiles

10.5 The Normal Distribution

10.1 Collecting Data

Statistical Terms

Statistics is the _____.

Data are the _____.

A **population** is the _____.

A **population parameter** is a number that describes _____.

A **census** is when data are collected from _____.

A **sample** is a _____.

A **sample statistic** is a number that describes _____.

▶ Watch and Work

Watch the video for the example in the software and follow along in the space provided.

Example 1: Identifying Populations and Samples

For the following scenarios, identify the population, sample, and whether the results represent a population parameter or a sample statistic.

- **a.** A school board recently read a report indicating that the time teens spend playing video games is increasing dramatically. To test the report, they surveyed local high school students. Of the 250 students surveyed, 58% indicated that they played video games at least 4 hours a day.

- **b.** A local youth group interviewed 800 adults across a southern state about their views on teenagers' use of cell phones. The resulting report stated that approximately 58% of the adults in the southern state were in favor of limiting teenagers' use of cell phones.

Solution

✏️ Now You Try It!

Use the space provided to work out the solution to the next example.

Example A: Identifying Populations and Samples

For the following scenarios, identify the population, sample, and whether the results represent a population parameter or a sample statistic.

a. A national non-profit organization interviewed 1000 children in a certain state and found that 43% of children in the state drink water with their lunch.

b. A study prompted by the increase of cell phone use looked at hospital records across the country. Of the 300 hospitals that released their records, 67% indicated an increase in ER visits caused by people distracted by cell phones.

Solution

Sampling Techniques

Random Sample: A sample where every member _____

_____ .

Systematic Sample: A sample where _____ .

Convenience Sample: A sample that is selected because it is _____

_____ .

Stratified Sample: A sample in which the population is _____

_____ and then a sample is taken from each group.

Cluster Sample: A sample in which the population is _____

_____

_____ .

10.1 Exercises

Concept Check

True/False. Determine whether each statement is true or false. If a statement is false, explain how it can be changed so the statement is true. (**Note:** There may be more than one acceptable change.)

1. A census collects data from every member of a sample.

2. A sample statistic gives information about the entire population.

3. In a study that interviews all high school students at the local school to find out what percentage of American teenagers have an iPhone, American teenagers are the population.

4. Asking everyone in your homeroom class what type of soda is their favorite is an example of stratified sampling.

Practice

5. Four scenarios of statistical studies are given below. Decide which study uses a sample statistic.

- 47% of American households own at least one dog.

- The National Institute of Pediatric Dentistry reports that 42% of children ages 2-11 have had at least one cavity.

- 1000 of the 1883 participants in last year's Ironman in Kona, Hawaii were interviewed. 68% of those interviewed said they became interested in triathlons after first competing in marathons.

- An estimated 15% of Americans don't use the internet.

6. Identify the population, the sample, and any population parameters or sample statistics in the given scenario.

Fit and Healthy magazine surveyed 1689 people. The resulting report stated that an estimated 56% of the U.S. population are self-conscious about their weight.

7. Identify the sampling technique used in the given scenario.

A local news agency is interested in studying unemployment rates across the state. The researchers will select several state counties at random, and then collect all relevant data from each county's unemployment office.

8. Choose the appropriate sampling method for the following scenario. Be sure to consider potential biases that the researcher might need to consider.

An EPA contractor needs to test the concentration of a substance in ten samples of the ground water. There are fifty possible locations to test, with no natural groupings or obvious differences.

Writing & Thinking

9. Suppose you are interested in what percentage of high school graduates own a pet. Describe a possible polling method you could use for each of the five sampling techniques. Which technique do you think would result in the most representative sample? What are some pros and cons for each technique?

10.2 Organizing and Displaying Data

A frequency distribution is the simplest way to view each observed _____ and know _____ .

A frequency distribution that is created after splitting the data into classes is called a _____ _____ .

When creating a grouped frequency distribution, it is important that the data is split so that all the classes are _____ . The number of data points in a class is called the _____ . The class width must be _____ for all classes and the classes _____ .

▶ Watch and Work

Watch the video for the example in the software and follow along in the space provided.

Example 3: Constructing a Grouped Frequency Distribution

The exam scores from two sections of a statistics course are recorded here.

| 85 | 78 | 82 | 79 | 84 | 88 | 71 | 73 | 89 | 77 | 84 | 94 | 98 | 86 | 77 | 79 | 73 | 72 | 89 | 88 |
| 86 | 85 | 79 | 76 | 81 | 80 | 92 | 75 | 88 | 82 | 81 | 91 | 88 | 93 | 92 | 79 | 80 | 78 | 82 | 96 |

Construct a grouped frequency distribution of the data collected with a first class of 71–76.

Solution

Chapter 10 Statistical Pathways: Introduction to Statistics

✏ Now You Try It!

Use the space provided to work out the solution to the next example.

Example A: Constructing a Grouped Frequency Distribution

The exam scores from two sections of a statistics course are recorded here.

| 94 | 100 | 94 | 71 | 92 | 69 | 63 | 96 | 69 | 91 |
| 79 | 93 | 82 | 79 | 100 | 98 | 66 | 67 | 73 | 74 |

Construct a grouped frequency distribution of the data collected with a first class of 61– 68.

Solution

A histogram is a special form of a _____ used to display the frequency distribution of numerical classes. In a histogram, _____.

A _____ is a type of line graph used to display the

_____.

A stem-and-leaf plot is a

_____.

10.2 Exercises

Concept Check

True/False. Determine whether each statement is true or false. If a statement is false, explain how it can be changed so the statement is true. (**Note:** There may be more than one acceptable change.)

1. The information in a frequency histogram is displayed in a two-column table where the first column describes the data values and the second column indicates the frequency of each of those data values.

2. Data values in frequency distributions can be names or labels (often referred to as categories) or they can be numerical.

3. If you are making a grouped frequency distribution for data values that range from 18 to 34, and the first class is 15–18, then the next class should be 18–21.

4. If a grouped frequency distribution has a class of 87–94, the number 87 is a lower class limit and the number 94 is an upper class limit.

Practice

5. The grades on the second statistics test for Mr. Montgomery's class are in the following list. Complete the frequency distribution table for the grades.

C, F, B, C, F, F, A, C, C, C, C, F, C, D, B, F, D, D, D, F, F

Grade	Frequency
A	
B	
C	
D	
F	

6. At a state fair, one game involves guessing the number of candies in a plastic jar. The following data represents the guesses that people made during one hour at the state fair. Complete the frequency table for this data.

1138, 1147, 1278, 1142, 1286, 1328, 1327, 1384, 1258, 1177, 1361, 1121, 1411, 1215, 1328, 1349

Determine the frequency of each class in the table shown.

7. The following histogram represents the distribution of scores on a ten point quiz.

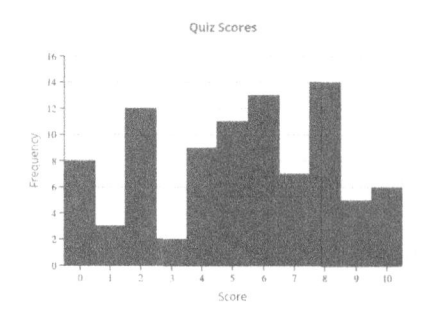

   a. Which score has the highest frequency?

   b. What is the frequency corresponding to a score of 10?

   c. What is the total number of people who made a score between 4 and 6 inclusive?

8. The following data represent the times in minutes required for 18 co-workers to commute to work. Use the given data to determine the stems for this stem-and-leaf plot.

57 47 43 36 21 26

42 31 46 21 35 32

45 31 40 32 38 22

Commute Times in Minutes	
Stem	Leaves
2	
3	
4	
5	

9. The following stem-and-leaf plot represents the distribution of weights for a group of people.

Stem	Leaves
11	2 4 5 5 6 6
12	1 3 3 4 5
13	4 6 6 7
14	2 4 4 7 7
15	0 4 6 6 7 9 9
16	6 6 7 8
17	2 4 6 6 9
18	5 7
19	0 1 5 9

a. What is the weight of the lightest person in the group?

b. How many people weigh in the range from 140 to 180 inclusive?

c. What is the weight of the heaviest person in the range 180 to 189 inclusive?

10. Which of the statements about the following graph is true?

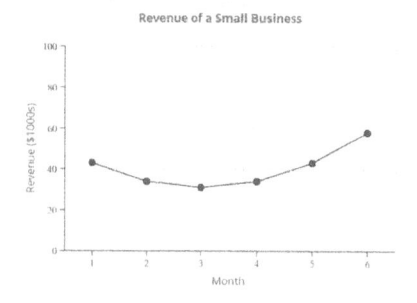

a. The *y*-scale is too large for the given data.

b. The *y*-scale is too small for the given data.

c. The *y*-scale is appropriate for the given data.

Writing and Thinking

11. Explain how to convert a frequency distribution to a grouped frequency distribution.

10.3 Measures of Center

Mean

The **mean** or **arithmetic average** is the _____

_____ .

The formula for mean is

$$\bar{x} = \underline{\hspace{3cm}},$$

where n is _____ .

More formally, the mean formula can be written using summation notation as _____ , where the Greek letter sigma, _____ , is used to denote the _____ .

How to Determine the Median

1. Arrange the data in _____ order.

2. a. If there is an odd number of data values, _____ _____ .

   b. If there is an even number of data values, _____ _____ .

Note: When there is an _____ of data points in a set, unless _____ , the median will not be a value in the data set.

Mode

The **mode** of a data set is _____. If all data values occur the same number of times, then there is _____.

A data set is _____ if there is only one data value that occurs most frequently.

A data set is **bimodal** if _____
_____.

If there are more than two data values that qualify to be the mode, then we say the data set is _____.

A **weighted mean** is one in which
_____.

▶ Watch and Work

Watch the video for the example in the software and follow along in the space provided.

Example 9: Finding the Weighted Mean

Let's assume in your mathematics class there are three categories that make up your grade: homework, quizzes, and exams. Your homework grade is 95%, your quiz grade is 85%, and your exam grade is 87%.

a. Find the mean of the three categories.

b. Your teacher assigns different weights to each category. Homework is 10% of your final grade, quizzes are 30%, and exams are 60%. Find your grade using the weighted mean.

Solution

10.3 Measures of Center

✏️ Now You Try It!

Use the space provided to work out the solution to the next example.

Example A: Finding the Weighted Mean

Let's assume in your statistics class there are three categories that make up your grade: homework, quizzes, and exams. Your homework grade is 92%, your quiz grade is 87%, and your exam grade is 82%.

   a. Find the arithmetic mean of the three categories.

   b. Your teacher assigns different weights to each category. Homework is 20% of your final grade, quizzes are 30%, and exams are 50%. Find your grade using the weighted mean.

Solution

10.3 Exercises

Concept Check

True/False. Determine whether each statement is true or false. If a statement is false, explain how it can be changed so the statement is true. (**Note:** There may be more than one acceptable change.)

1. The average of a set of numbers is also called the arithmetic average or mean.

2. To find the mean of a set of 10 numbers, you would find the sum of all the numbers and divide by the smallest value.

3. The middle value in a ranked set of data is called the median.

4. If all data values occur the same number of times, then each data value is a mode.

Practice

5. Consider the following data.

12, 13, 14, 3, 2, 10

  a. Determine the mean of the given data.

  b. Determine the median of the given data.

  c. Determine if the data set is unimodal, bimodal, multimodal, or has no mode. Identify the mode(s), if any exist.

6. Consider the following data.

12, 5, 2, 9, 11, 8, 2

  a. Determine the mean of the given data.

  b. Determine the median of the given data.

  c. Determine if the data set is unimodal, bimodal, multimodal, or has no mode. Identify the mode(s), if any exist.

7. Given the following weights and data values, calculate the mean. Round your answer to three decimal places.

Weights	1.97	1.92	1.18	5.65	4.65	3.09	1.51	6.62
Data Values	4	5	7	7	8	3	2	5

Applications

Solve.

8. Calculate the GPA of a student with the following grades: C (17 hours), A (21 hours), D (13 hours). Note that an A is equivalent to 4.0, a B is equivalent to a 3.0, a C is equivalent to a 2.0, a D is equivalent to a 1.0, and an F is equivalent to a 0. Round your answer to two decimal places.

9. Twenty students in Highpoint Middle School's eighth grade class are collecting signatures on a petition for the cafeteria to offer "Taco Tuesday" twice a month. Use the following frequency distribution of the number of signatures collected by the students to determine the following for the given data concerning the numbers of signatures obtained so far.

Number of Signatures	Frequency
4	2
5	2
6	5
7	3
8	4
9	3
10	1

a. the mean,

b. the median, and

c. the mode

Writing & Thinking

10. Give an example of when the mode would be the best measure of central tendency to use.

10.4 Measures of Dispersion and Percentiles

> **Range**
>
> The **range** is the _____
>
> _____ .

▶ Watch and Work

Watch the video for the example in the software and follow along in the space provided.

Example 2: Finding the Range

The grades on a recent statistics exam are listed below. Calculate the range.

$$85 \quad 88 \quad 82 \quad 84 \quad 89 \quad 81 \quad 90 \quad 63 \quad 87 \quad 86$$

Solution

✏ Now You Try It!

Use the space provided to work out the solution to the next example.

Example A: Finding the Range

The grades on a recent mathematics exam are listed below. Calculate the range.

$$72 \quad 81 \quad 94 \quad 78 \quad 83 \quad 81 \quad 98 \quad 87 \quad 68 \quad 92$$

Solution

10.4 Measures of Dispersion and Percentiles

Sample Standard Deviation

The **sample standard deviation** indicates _____
_____.

The formula for standard deviation for a sample is

$$,$$

where x represents each data value, $\bar{x}$ is the sample mean, and n is the sample size. Recall that the Greek letter sigma, Σ, is used to denote summation.

A _____ describes the position of a specific data value in a data set compared to the rest of the values in the set.

_____ divide the data set up into 100 equal parts and indicate approximately what percentage of the data lies at or below the data value.

Quartiles

First Quartile $= Q_1 =$ _____, that is, _____ of the data are at or below this data value.

Second Quartile $= Q_2 =$ _____, that is, _____ of the data are at or below this data value.

Third Quartile $= Q_3 =$ _____, that is, _____ of the data are at or below this data value.

It is important to note that, by definition, Q_2 is the same as the _____.

10.4 Exercises

Concept Check

True/False. Determine whether each statement is true or false. If a statement is false, explain how it can be changed so the statement is true. (**Note:** There may be more than one acceptable change.)

1. A measure of dispersion that takes only two data values into consideration is the standard deviation.

2. The median is the second quartile.

3. Quartiles divide a data set into eight equal parts.

4. A large the standard deviation indicates that the data values are spread out from the mean.

Practice

5. Find the range of the following data set. Assume the data set is a sample.

28, 23, 23, 24, 18, 18, 25, 17, 11, 30, 15

6. Find the standard deviation of the following data set. Assume the data set is a sample. Round your answer to the nearest hundredth, if necessary.

20, 40, 16, 22, 35, 23, 23, 29, 28, 28, 38

7. The data set represents the test scores for 16 students in a class on their most recent test. Use the data provided to find the quartiles.

51, 51, 52, 53, 55, 59, 60, 64, 65, 72, 77, 78, 84, 84, 84, 85

   a. Find the *second* quartile.

   b. Find the *first* quartile.

   c. Find the *third* quartile.

Applications

Solve.

8. A researcher randomly purchases several different kits of a popular building toy. The following table shows the number of pieces in each kit in the sample. Find the range of the data.

Building Toy Pieces	
151	36
256	341
341	36
95	287
256	100
161	453

9. A researcher randomly purchases several different kits of a popular building toy. The following table shows the number of pieces in each kit in the sample. Find the standard deviation of the data. Round your answer to the nearest hundredth, if necessary.

Building Toy Pieces	
155	40
260	345
345	40
99	291
260	104
165	465

10. A graduate school only accepts applicants who score in the top 15% on the GRE. Which score falls within the top 15%?

    a. 20th percentile

    b. 90th percentile

    c. 70th percentile

d. 10th percentile

Writing & Thinking

11. Give an example of two data sets where the range is the same but the standard deviation is different.

10.5 The Normal Distribution

A _____ is a symmetrical, bell-shaped distribution. Most of the data points in a normal distribution are _____. All normal distributions are defined by a _____, which acts as the line of symmetry, and a _____, which determines how the data spreads out from the mean.

Characteristics of the Normal Distribution

1. It is _____.

2. The mean, median, and mode are all _____.

3. The total area under the curve is _____.

4. The distribution is completely defined by its _____.

The Empirical Rule

- Approximately 68% of the data values fall within _____ of the mean.

- Approximately 95% of the data values fall within _____ of the mean.

- Approximately 99.7% of the data values fall within _____ of the mean.

z-Score

The z-score tells us

_____.

$$z = \underline{\hspace{3in}}$$

More formally with symbols, the formula is

$$z = \underline{\hspace{1in}}.$$

for populations where μ is the mean and σ is standard deviation.

▶ Watch and Work

Watch the video for the example in the software and follow along in the space provided.

Example 5: z-Score Application

Sally and Janet each took a standardized test for their respective areas of study in preparation to apply to graduate school. Although they took different tests, they want to compare how well they performed on the tests. Sally scored 1280, and her test had a mean of 1000 with a standard deviation of 140. Janet scored 1440, and her test had a mean of 1200 with a standard deviation of 160. The scores for both tests are approximately normally distributed. Who performed better on her test, Sally or Janet?

Solution

✏️ Now You Try It!

Use the space provided to work out the solution to the next example.

Example A: *z*-score Application

James and Eli each took a standardized test for their respective areas of study in preparation to apply to graduate school. Although they took different tests, they want to compare how well they performed on the tests. James scored 1420, and his test had a mean of 1250 with a standard deviation of 100. Eli scored 1380, and his test had a mean of 1200 with a standard deviation of 120. The scores for both tests are approximately normally distributed. Who performed better on his test, James or Eli?

Solution

10.5 Exercises

Concept Check

True/False. Determine whether each statement is true or false. If a statement is false, explain how it can be changed so the statement is true. (**Note:** There may be more than one acceptable change.)

1. A positive *z*-score indicates that the data value is greater than the mean.

2. It is not possible for a *z*-score to be 0.

3. The normal distribution is bimodal.

4. Changes in the mean and standard deviation can make normal distributions look different from one another while retaining the same general shape.

Practice

5. Suppose that grade point averages of undergraduate students at one university have a bell-shaped distribution with a mean of 2.52 and a standard deviation of 0.37. Using the empirical rule, what percentage of the students have grade point averages that are between 2.15 and 2.89?

6. Suppose that IQ scores have a bell-shaped distribution with a mean of 100 and a standard deviation of 15. Using the empirical rule, what percentage of IQ scores are greater than 130? Please do not round your answer.

7. Suppose that shoe sizes of American women have a bell-shaped distribution with a mean of 8.2 and a standard deviation of 1.49. Using the empirical rule, what percentage of American women have shoe sizes that are less than 6.71? Please do not round your answer.

8. Suppose that IQ scores have a bell-shaped distribution with a mean of 101 and a standard deviation of 17. Describe where the highest and lowest 0.3% of IQ scores lie.

9. Calculate the standard score of the given x value, $x = 54$, where $\mu = 53.1$, $\sigma = 7.1$. Round your answer to two decimal places.

10. Scores on a test have a mean of 60 and a standard deviation of 9. Tony has a score of 71. Convert Tony's score to a z-score, rounded to the nearest hundredth.

11. The rainfall in a town last year measured 75 inches. If the annual rainfall has a mean of 47.41 inches and a standard deviation of 9.56 inches, how many standard deviations away from the mean was the rainfall last year? Round your answer to two decimal places.

Writing & Thinking

12. Describe how the distribution curves of data sets with the same mean but different standard deviations would look different.

Chapter A
Appendix

A.1 Matrices and Basic Matrix Operations

A.1 Matrices and Basic Matrix Operations

A _____ is called a **matrix** (plural **matrices**).

The dimensions of a matrix are determined by the _____ and _____ _____ and are written in the form _____.

Addition with Matrices

Two matrices A and B can be added as long as they _____.

Each entry in matrix $A + B$ is found by _____ in matrix A and matrix B.

Subtraction with Matrices

Two matrices A and B can be subtracted as long as they _____.

Each entry in matrix $A - B$ is found by _____ in matrix A and matrix B.

Product of a Scalar and a Matrix

To find the product of a scalar and a matrix, _____
_____.

Multiplication with Matrices

If matrix A has dimensions $n \times p$ and matrix B has dimensions $p \times m$, then the product of matrices A and B (denoted AB) is an _____ matrix. The entry in the ith row and j^{th} column of AB is found by multiplying _____
_____.

▶ Watch and Work

Watch the video for the example in the software and follow along in the space provided.

Example 6: Multiplying Two Matrices

Find the product AB using the given matrices.

$$A = \begin{bmatrix} -4 & -8 \\ 2 & 0 \\ -1 & 3 \end{bmatrix} \quad \text{and} \quad B = \begin{bmatrix} 8 & -6 \\ 5 & 1 \end{bmatrix}$$

Solution

✏️ Now You Try It!

Use the space provided to work out the solution to the next example.

Example A: Multiplying Two Matrices

Find the product AB using the given matrices.

$$A = \begin{bmatrix} 3 & -5 \\ 4 & -1 \\ 0 & 9 \end{bmatrix} \text{ and } B = \begin{bmatrix} 1 & 7 \\ -8 & 2 \end{bmatrix}$$

Solution

A.1 Exercises

Concept Check

True/False. Determine whether each statement is true or false. If a statement is false, explain how it can be changed so the statement is true. (**Note:** There may be more than one acceptable change.)

1. The double subscript notation b_{43} is read "b four three".

2. When adding or subtracting two matrices, the matrices must have the same dimensions.

3. The dimensions of a matrix are written as the number of columns by the number of rows.

4. The number of columns in A need to match the number of rows in B in order for the product AB to be found.

Practice

5. Determine the dimensions of the given matrix

$$Q = \begin{bmatrix} 5 & 3 & -8 \\ 10 & 1 & 0 \\ -2 & 1 & -3 \end{bmatrix}$$

6. Determine the dimensions of the given matrix

$$R = \begin{bmatrix} -3 & -2 & 3 & -4 & 3 \\ -6 & 5 & 6 & 6 & 8 \end{bmatrix}$$

7. Perform the indicated operation, if possible.

$$A + B = \begin{bmatrix} 0 & -8 & 10 \\ 0 & 2 & 9 \end{bmatrix} + \begin{bmatrix} 6 & 4 & -2 \\ -7 & -7 & 6 \end{bmatrix}$$

8. Perform the indicated operation, if possible.

$$X + Y = \begin{bmatrix} 8 & 9 & 5 \\ 5 & 5 & -3 \\ 7 & 9 & 0 \end{bmatrix} + \begin{bmatrix} -1 & -5 & -9 \\ 10 & 3 & 0 \\ 6 & 6 & 3 \end{bmatrix}$$

9. Perform the indicated operation, if possible.

$$C - D = \begin{bmatrix} -3 & -3 \\ 2 & -1 \end{bmatrix} - \begin{bmatrix} -6 & 9 & -9 \\ 0 & -1 & -1 \end{bmatrix}$$

10. Perform the indicated operation, if possible.

$$W - V = \begin{bmatrix} -7 & 9 \\ -5 & 0 \\ 0 & 2 \\ -8 & 6 \end{bmatrix} - \begin{bmatrix} -1 & -10 \\ -8 & 5 \\ -1 & -1 \\ 1 & 8 \end{bmatrix}$$

11. Perform the indicated operation, if possible.

$$8B = 8 \begin{bmatrix} -7 & 7 \\ -4 & -3 \\ -3 & -6 \end{bmatrix}$$

12. Perform the indicated operation, if possible.

$$GH = \begin{bmatrix} 0 & -6 & 1 \end{bmatrix} \begin{bmatrix} 6 & 6 \\ 8 & -2 \\ 1 & 1 \end{bmatrix}$$

Math@Work

Basic Inventory Management 385

Hospitality Management: Preparing for a Dinner Service 387

Bookkeeper 389

Pediatric Nurse 391

Architecture 393

Statistician: Quality Control 395

Dental Assistant 397

Financial Advisor 399

Market Research Analyst 401

Astronomy 403

Math Education 405

Forensic Scientist 407

Other Careers in Mathematics 409

Math@Work
Basic Inventory Management

As an business manager you will need to evaluate the company's inventory several times per year. While evaluating the inventory, you will need to ensure that enough of each product will be in stock for future sales based on current inventory count, predicted sales, and product cost. Let's say that you check the inventory four times a year, or quarterly. You will be working with several people to get all of the information you need to make the proper decisions. You need the sales team to give you accurate predictions of how much product they expect to sell. You need the warehouse manager to keep an accurate count of how much of each product is currently in stock and how much of that stock has already been sold. You will also have to work with the product manufacturer to determine the cost to produce and ship the product to your company's warehouse. It's your job to look at this information, compare it, and decide what steps to take to make sure you have enough of each product in stock for sales needs. A wrong decision can potentially cost your company a lot of money.

Suppose you get the following reports: an inventory report of unsold products from the warehouse manager and the report on predicted sales for the next quarter (three months) from the sales team.

Unsold Products	
Item	Number in Stock
A	5025
B	150
C	975
D	2150

Predicted Sales	
Item	Expected Sales
A	4500
B	1625
C	1775
D	2000

Suppose the manufacturer gives you the following cost list for the production and shipment of different amounts of each inventory item.

Item	Amount	Cost	Amount	Cost	Amount	Cost
A	500	$875	1000	$1500	1500	$1875
B	500	$1500	1000	$2500	1500	$3375
C	500	$250	1000	$400	1500	$525
D	500	$2500	1000	$4250	1500	$5575

1. Which items and how much of each item do you need to purchase to make sure the inventory will cover the predicted sales?

2. If you purchase the amounts from Problem 1, how much will this cost the company?

3. By ordering the quantities you just calculated, you are ordering the minimum of each item to cover the expected sales. If the actual sales during the quarter are higher than expected, what might happen? How would you handle this situation?

4. Which skills covered in this chapter were necessary to help you make your decisions?

Math@Work

Hospitality Management: Preparing for a Dinner Service

As the manager of a restaurant you will need to make sure everything is in place for each meal service. This means that you need to predict and prepare for busy times, such as a Friday night dinner rush. To do this, you will need to obtain and analyze information to determine how much of each meal is typically ordered. After you estimate the number of meals that will be sold, you need to communicate to the chefs how much of each item they need to expect to prepare. An additional aspect of the job is to work with the kitchen staff to make sure you have enough ingredients in stock to last throughout the meal service.

You are given the following data, which is the sales records for the signature dishes during the previous four Friday night dinner services.

Week	Meal A	Meal B	Meal C	Meal D
1	30	42	28	20
2	35	38	30	26
3	32	34	26	26
4	30	32	28	22

Meal C is served with a risotto, a type of creamy rice. The chefs use the following recipe, which makes 6 servings of risotto, when they prepare Meal C. (**Note:** The abbreviation for tablespoon is T and the abbreviation for cup is c.)

$5\frac{1}{2}$ c chicken stock $2\frac{1}{3}$ T chopped shallots $\frac{1}{2}$ c red wine

$1\frac{1}{2}$ c rice   2 T chopped parsley $4\frac{3}{4}$ c thinly sliced mushrooms

2 T butter   2 T olive oil $\frac{1}{2}$ c Parmesan cheese

1. For the past four Friday night dinner services, what was the average number of each signature meal served? If the average isn't a whole number, explain why you would round this number either up or down.

2. Based on the average you obtained for Meal C, calculate how much of each ingredient your chefs will need to make the predicted amount of risotto.

3. The head chef reports the following partial inventory: $10\frac{3}{4}$ c rice, $15\frac{3}{4}$ c mushrooms, and 10 T shallots. Do you have enough of these three items in stock to prepare the predicted number of servings of risotto?

4. Which skills covered in this and the previous chapter helped you make your decisions?

Name: Date: **389**

Math@Work

Bookkeeper

As a bookkeeper, you will often receive bills and receipts for various purchases or expenses from employees of the company you work for. You will need to split the bill by expense code, assign costs according to customer, and reimburse an employee for their out-of-pocket spending. To do this you will need to know the company's reimbursement policies, the expense codes for different spending categories, and which costs fall into a particular expense category.

Suppose two employees from the sales department recently completed sales trips. Employee 1 flew out of state and visited two customers, Customer A and Customer B. This employee had a preapproved business meal with Customer B and was traveling for three days. Employee 2 drove out of state to visit Customer C. This employee stayed at a hotel for the night and then drove back the next day. The expenses for the two employees are as follows.

Employee 1	
Flight and Rental Car	$470.50
Hotel	$278.88
Meals	$110.56
Business Meal	$102.73
Presentation Materials	$54.86

Employee 2	
Miles Driven	578.5 miles
Fuel	$61.35
Hotel	$79.60
Meals	$53.23
Presentation Materials	$67.84

The expense categories used by your company to track spending are: Travel (includes hotel, flights, mileage, etc.), Meals (business), Meals (travel), and Supplies. Traveling employees are reimbursed up to $35 per day for meals while traveling and for all preapproved business meals. They also receive $0.565 per mile driven with their own car.

1. How much will you reimburse each employee for travel meals? Did either employee go over their allowed meal reimbursement amount?

2. What were the total expenses for each employee?

3. The company you work for keeps track of how much is spent on each customer. When a sales person visits multiple customers during one trip, the tracked costs are split between the customers. Fill in this table according to how much was spent on each customer for the different expense categories. (**Note:** For meals, only include the amount the employee was reimbursed.)

Expense	Customer A	Customer B	Customer C
Travel			
Meals (business)			
Meals (travel)			
Supplies			
Total			

Name: _____   Date: _____

Math@Work
Pediatric Nurse

As a pediatric nurse working in a hospital setting, you will be responsible for taking care of several patients during your work day. You will need to administer medications, set IVs, and check each patient's vital signs (such as temperature and blood pressure). While doctors prescribe the medications that nurses need to administer, it is important for nurses to double check the dosage amounts. Administering the incorrect amount of medication can be detrimental to the patient's health.

During your morning nursing round, you check in on three new male patients and obtain the following information.

	Patient A	Patient B	Patient C
Age	10	9	12
Weight (pounds)	81	68.5	112
Blood Pressure	97/58	100/59	116/73
Temperature (°F)	99.7	97.3	101.4
Medication	A	B	A

The following table shows the bottom of the range for abnormal blood pressure (BP) for boys. If either the numerator or the denominator of the blood pressure ratio is greater than or equal to the values in the chart, this can indicate a stage of hypertension.

Abnormal Blood Pressure for Boys by Age	
	Systolic BP / Diastolic BP
Age 9	109/72
Age 10	111/73
Age 11	113/74
Age 12	115/74

Source: http://www.nhlbi.nih.gov/health/public/heart/hbp/bp_child_pocket/bp_child_pocket.pdf

Medication Directions	
Medication	Dosage Rate
A	40 mg per 10 pounds
B	55 mg per 10 pounds

1. Do any of the patients have a blood pressure which may indicate they have hypertension? If yes, which patient(s)?

2. Use proportions to determine the amount of medication that should be administered to each patient based on weight. Round to the nearest 10 pounds before calculating.

3. The average body temperature is 98.6 degrees Fahrenheit. You are supposed to alert the doctor on duty if any of the patients have a temperature 2.5 degrees higher than average. For which patients would you alert a doctor?

4. Which skills covered in this chapter and the previous chapters were necessary to help you make your decisions?

Name: Date: 393

Math@Work

Architecture

As a project architect, you will be part of a team that creates detailed drawings of the project that will be used during the construction phase. It will be your job to ensure that the project will meet guidelines given to you by your company, such as square footage requirements and budget constraints. You will also need to meet the design requirements requested by the client.

Suppose you are part of a team that is designing an apartment building. You are given the task to create the floor plan for an apartment unit with two bedrooms and one bathroom. The apartment management company that has contracted your company to do the project has several requirements for this specific apartment unit.

1. One bedroom is the "master bedroom" and must have at least 60 square feet more than the other bedroom.
2. All walls must intersect or touch at 90 degree angles.
3. The kitchen must have an area of no more than 110 square feet.
4. The apartment must be between 1000 square feet and 1050 square feet.

A preliminary sketch of the apartment is shown here.

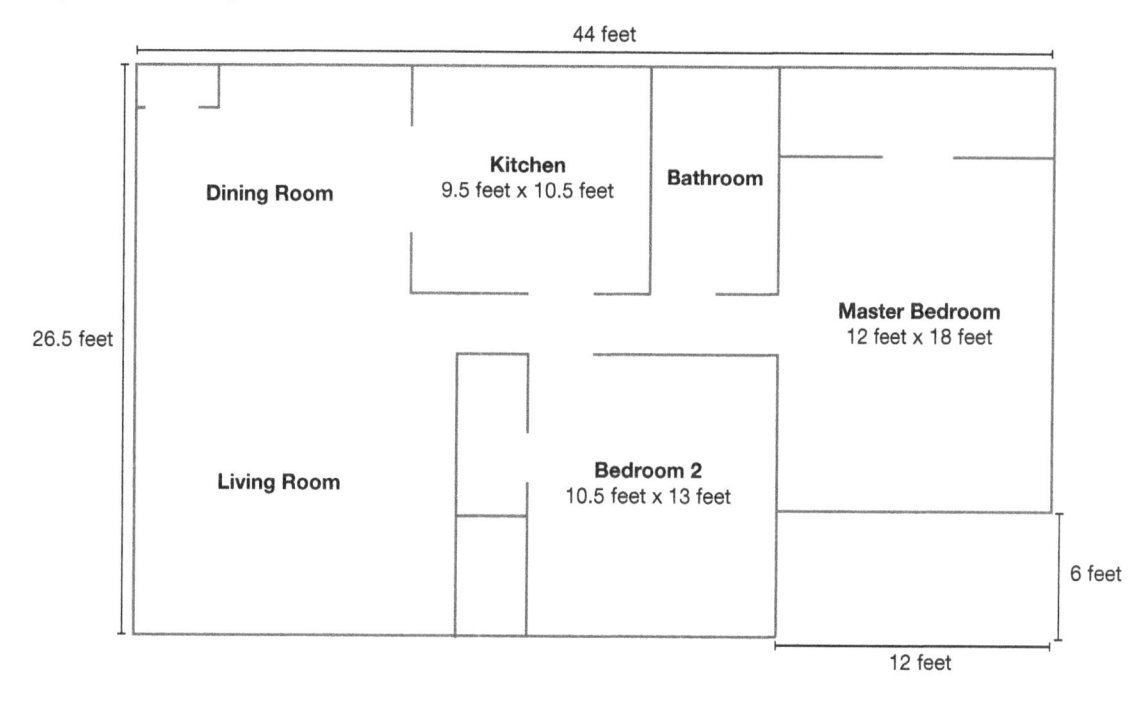

[Floor plan diagram: Overall dimensions 44 feet wide by 26.5 feet tall. Rooms shown: Dining Room, Kitchen (9.5 feet x 10.5 feet), Bathroom, Master Bedroom (12 feet x 18 feet), Living Room, Bedroom 2 (10.5 feet x 13 feet). A 12 feet by 6 feet notch is cut out of the lower right corner.]

1. Does the apartment have the required total square footage that was requested? Is it over or under the total required?

2. Does the apartment blueprint meet the other requirements given by the client? If not, what does not meet the requirements?

3. For this specific apartment unit, the total construction cost per square foot is estimated to be $75.75. Approximately how much will it cost to construct each two-bedroom apartment based on the floor plan?

Math@Work
Statistician: Quality Control

Suppose you are a statistician working in the quality control department of a company that manufactures the hardware sold in kits to assemble book shelves, TV stands, and other ready-to-assemble furniture pieces. There are three machines that produce a particular screw and each machine is sampled every hour. A measurement of the screw length is determined with a micrometer, which is a device used to make highly precise measurements. The screw is supposed to be 3 inches in length and can vary from this measurement by no more than 0.1 inches or it will not fit properly into the furniture. The following table shows the screw length measurements (in inches) taken each hour from each machine throughout the day. The screw length data from each machine has also been plotted

Screw Length Measurements (in inches)			
Sample Time	Machine A	Machine B	Machine C
8 a.m.	2.98	2.92	2.99
9 a.m.	3.00	2.94	3.00
10 a.m.	3.02	2.97	3.01
11 a.m.	2.99	2.96	3.03
12 p.m.	3.01	2.94	3.05
1 p.m.	3.00	2.95	3.04
2 p.m.	2.97	2.93	3.06
3 p.m.	2.99	2.92	3.08
Mean			
Range			

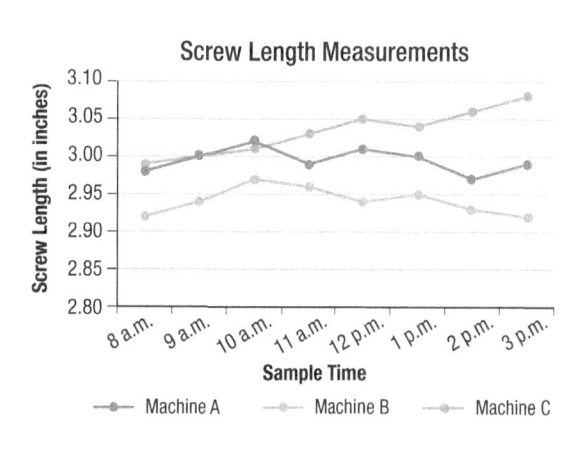

1. Calculate the mean and range of the data for each machine and place them in the bottom two rows of the table.

2. If the screw length can vary from 3 inches by no more than 0.1 inch (plus or minus), what are the lowest and highest values for length that will be acceptable? Place a horizontal line on the graph at each of these values on the vertical axis. These are the tolerance or specification limits for screw length.

3. Have any of the three machines produced an unacceptable part today? Are any of the machines close to making a bad part? If so, which one(s)?

4. Look at the graph and the means from the table that show the average screw length produced by each machine. Draw a bold horizontal line on the graph at 3 to emphasize the target length. Do all the machines appear to be making parts that vary randomly around the target of 3 inches?

5. Look at the range values from the table. Do any of the machines appear to have more variability in the length measurements than the others?

6. In your opinion, which machine is performing best? Would you recommend that any adjustments be made to any of the machines? If so, which one(s) and why?

Math@Work
Dental Assistant

As a dental assistant, your job duties will vary depending on where you work. Suppose you work in a dental office where you assist with dental procedures and managing patients' accounts. When a patient arrives for their appointment, you will need to review their chart and make sure they are up to date on preventive care, such as X-rays and cleanings. When the patient leaves, you will need to fill out an invoice to determine how much to charge the patient for their visit.

Dental patients generally have a new X-ray taken yearly. Cleanings are performed every 6 months, although some patients have their teeth cleaned more often. The following table shows the date of the last X-ray and cleaning for three patients that are visiting the office today. (**Note:** All dates are within the past year.)

Patient Histories		
Patient	**Last X–ray**	**Last Cleaning**
A	April 15	October 20
B	June 6	January 12
C	October 27	October 27

During Patient A's visit, she received a fluoride treatment and a cleaning. Patient A has no dental insurance. During Patient B's visit, he received a filling on one surface of a tooth. Patient B has dental insurance which pays for 60% of the cost of fillings. During Patient C's visit, he had a cleaning, a filling on one surface of a tooth, and a filling on two surfaces of another tooth. Patient C has dental insurance which covers the full cost of cleanings and 50% of the cost of fillings.

Fee Schedule	
Procedure	**Cost**
Cleaning	$95
Fluoride treatment	$35
Filling, One surface	$175
Filling, Two surfaces	$235
X–ray, Panoramic	$110

1. Using today's date, determine which of the three patients are due for a dental cleaning in the next two months?

2. Using today's date, determine which of the patients will require a new set of X-rays during this visit.

3. Determine the amount each patient will be charged for their visit (without insurance). Don't forget to include the cost of any X-rays that are due during the visit.

4. Use the insurance information to determine the amount that each patient will pay out-of-pocket at the end of their visit.

Math@Work
Financial Advisor

As a financial advisor working with a new client, you must first determine how much money your client has to invest. The client may have a lump sum that they have saved or inherited, or they may wish to contribute an amount monthly from their current salary. In the latter case, you must then have the client do a detailed budget, so that you can determine a reasonable amount that the client can afford to set aside on a monthly basis for investment.

The second piece of information necessary when dealing with a new client is determining how much risk-tolerance they have. If the client is young or has a lot of money to invest, they may be willing to take more risk and invest in more aggressive, higher interest-earning funds. If the client is older and close to retirement, or has little money to invest, they may prefer less-aggressive investments where they are essentially guaranteed a certain rate of return. The range of possible investments that would suit each client's needs and goals are determined using a survey of risk-tolerance.

Suppose you have a client who has a total of $25,000 to invest. You determine that there are two investment funds that meet the client's investment preferences. One option is an aggressive fund that earns an average of 12% interest and the other is a more moderate fund that earns an average of 5% interest. The client desires to earn $2300 this year from these investments.

Investment Type	Principal Invested	·	Interest Rate	=	Interest Earned
Aggressive Fund	x				
Moderate Fund					

To determine the amount of interest earned you know to use the table above and the formula $I = Prt$, where I is the interest earned, P is the principal or amount invested, r is the average rate of return, and t is the length of time invested. Since the initial investment will last one year, $t = 1$.

1. Fill in the table with the known information. If x is the amount invested in the aggressive fund and the total amount to be invested is $25,000, create an expression involving x for the amount that will be left to invest in the moderate fund. Place this expression in the appropriate cell of the table.

2. Determine an expression in x for the interest earned on each investment type by multiplying the principal by the interest rate.

3. Determine the amount invested in each fund by setting up an equation using the expressions in column four and the fact that the client desires to earn $2300 from the interest earned on both investments.

4. Verify that the investment amounts calculated for each fund in the previous step are correct by calculating the actual interest earned in a year for each and making sure they sum to $2300.

5. Why would you not advise your client to invest all their money in the fund earning 12% interest, after all, it has the highest average interest rate?

Math@Work
Market Research Analyst

As a market research analyst, you may work alone at a computer, collecting and analyzing data, and preparing reports. You may also work as part of a team or work directly with the public to collect information and data. Either way, a market research analyst must have strong math and analytical skills and be very detail-oriented. They must have strong critical-thinking skills to assess large amounts of information and be able to develop a marketing strategy for the company. They must also possess good communication skills in order to interpret their research findings and be able to present their results to clients.

Suppose you work for a shoe manufacturer who wants to produce a new type of lightweight basketball sneaker similar to a product a competitor recently released into the market. You have gathered some sales data on the competitor in order to determine if this venture would be worthwhile, which is shown in the table below. To begin your analysis, you create a scatter plot of the data to see the sales trend. (A scatter plot is a graph made by plotting ordered pairs in a coordinate plane in order to show the relationship between two variables.) You determine that the *x*-axis will represent the number of weeks after the competitors new sneaker went on the market and the *y*-axis will represent the amount of sales in thousands of dollars.

Number of Weeks x	Sales (in 1000s) y
3	15
6	22
9	28
12	35
15	43

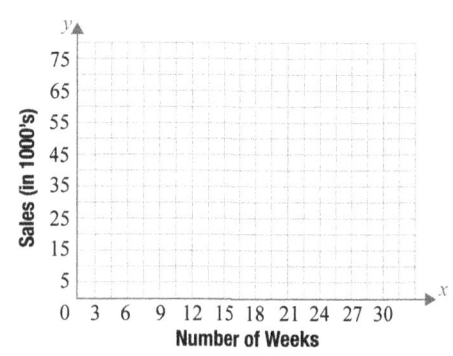

1. Create a scatter plot of the sales data by plotting the ordered pairs in the table on the coordinate plane. Does the data on the graph appear to follow a linear pattern? If so, sketch a line that you feel would "best" fit this set of data. (A market research analyst would typically use computer software to perform a technique called regression analysis to fit a "best" line to this data.)

2. Using the ordered pairs corresponding to weeks 9 and 15, find the equation of a line running through these two data points.

3. Interpret the value calculated for the slope of the equation in Problem 2 as a rate of change in the context of the problem. Write a complete sentence.

4. If you assume that the sales trend in sneaker sales follows the model determined by the linear equation in Problem 2, predict the sneaker sales in 6 months. Use the approximation that 1 month is equal to 4 weeks.

5. Give at least two reasons why the assumption made in Problem 4 may be invalid?

Math@Work
Astronomy

Astronomy is the study of celestial bodies, such as planets, asteroids, and stars. While you work in the field of astronomy, you will use knowledge and skills from several other fields, such as mathematics, physics, and chemistry. An important tool of astronomers is the telescope. Several powerful telescopes are housed in observatories around the world. One of the many things astronomers use observatories for is discovering new celestial objects such as a near-Earth object (NEO). NEOs are comets, asteroids, and meteoroids that orbit the sun and cross the orbital path of Earth. The danger presented by NEOs is that they may strike the Earth and result in global catastrophic damage. (**Note:** The National Aeronautics and Space Administration (NASA) keeps track of all NEOs which are a potential threat at the website http://neo.jpl.nasa.gov/risk/)

For an asteroid to be classified as an NEO, the asteroid must have an orbit that partially lies within 0.983 and 1.3 astronomical units (AU) from the sun, where 1 AU is the furthest distance from the Earth to the sun, approximately 9.3×10^7 miles.

Near-Earth Object Distance			
	Minimum		**Maximum**
Distance in AU	0.983 AU	1 AU	1.3 AU
Distance in Miles		9.3×10^7 miles	

Suppose you discover three asteroids that you suspect may be NEOs. You perform some calculations and come up with the following facts. The furthest that Asteroid A is ever from the sun is 81,958,000 miles. The closest Asteroid B is ever to the sun is 12,529,000 miles. The closest Asteroid C is ever to the sun is 92,595,000 miles.

1. To determine if any of the asteroids pass within the range to be classified as an NEO, fill in the missing values from the table.

2. Based on the measurements from Problem 1, do any of the three asteroids qualify as an NEO?

There are two scales that astronomers use to explain the potential danger of NEOs. The Torino Scale is a scale from 0 to 10 that indicates the chance that an object will collide with the Earth. A rating of 0 means there is an extremely small chance of a collision and a 10 indicates that a collision is certain to happen. The Palermo Technical Impact Hazard Scale is used to rate the potential impact hazard of an NEO. If the rating is less than −2, the object poses a very minor threat with no drastic consequences if the object hits the Earth. If the rating is between −2 and 0, then the object should be closely monitored as it could cause serious damage.

Go to the NASA website http://neo.jpl.nasa.gov/risk/ to answer the following questions.

3. Does any NEO have a Torino Scale rating higher than 0? If so, what is the object's designation (or name) and during which year range could a potential impact occur?

4. Which NEO has the highest Palermo Scale rating? During which year range could a potential impact occur?

Name: Date:

Math@Work
Math Education

As a math instructor at a public high school, your day will be spent preparing class lectures, grading assignments and tests, and teaching students with a wide variety of backgrounds. While teaching math, it is your job to explain the concepts and skills of math in a variety of ways to help students learn and understand the material. As a result, a solid understanding of math and strong communication skills are very important. Teaching math is a challenge and being able to understand the reasons that students struggle with math and empathize with these students is a critical aspect of the job.

Suppose that the next topics you plan to teach to your algebra students involve finding the greatest common factor and factoring by grouping. To teach these skills, you will need to plan how much material to cover each day, choose examples to walk through during the lecture, and assign in-class work and homework. You decide to spend the first day on this topic explaining how to find the greatest common factor of a list of integers.

1. It is usually easier to teach a group of students a new topic by initially showing them a single method. If a student has difficulty with that method, then showing the student an alternative method can be helpful. Which method for finding the greatest common factor would you teach to the class during the class lecture?

2. On a separate piece of paper, sketch out a short lecture on finding the greatest common factor of a list of integers. Be sure to include examples that range from easy to difficult.

3. While the class is working on an in-class assignment, you find that a student is having trouble following the method that you taught to the entire class. Describe an alternative method that you could show the student.

4. From your experience with learning how to find the greatest common factor of a list of integers, what do you think are some areas that might confuse students and cause them to struggle while learning this topic? Explain how understanding the areas that might cause confusion can help you become a better teacher.

Math@Work
Forensic Scientist

As a forensic scientist, you will work as part of a team to investigate the evidence from a crime scene. Every case you encounter will be unique and the work may be intense. Communication is especially important because you will need to be clear and honest about your findings and your conclusions. A suspect's freedom may depend on the conclusions your team draws from the evidence.

Suppose the most recent case that you are involved in is a hit-and-run accident. A body was found at the side of the road with skid marks nearby. The police are unsure if the cause of death of the victim was vehicular homicide. Among the case description, the following information is provided to you.

Accident Report	
Date:	June 14
Time:	9:30 pm
Climate:	55 degrees Fahrenheit, partly cloudy, dry
Description of crime scene:	
Victim was found at the side of a road. Body temperature upon arrival is 84.9 °F. Posted speed limit is 30 mph. Road is concrete. Conditions are dry. Skid marks near the body are 88 feet in length.	

Known formulas and data:

A body will cool at a rate of 2.7 °F per hour until the body temperature matches the temperature of the environment. Average human body temperature is 98.6 °F.

Impact Speed and Risk of Death	
Impact Speed	**Risk of Death**
23 mph	10%
32 mph	25%
42 mph	50%
58 mph	90%
Source: 2011 AAA Foundation for Traffic Safety "Impact Speed and Pedestrian's Risk of Severe Injury or Death"	

Braking distance is calculated using the formula $\frac{s}{\sqrt{l}} = k$, where s is the initial speed of the vehicle in mph, l is the length of the skid marks in feet, and k is a constant that depends on driving conditions. Based on the driving conditions on that road for the last 12 hours, $k = \sqrt{20}$.

1. Based on the length of the skid marks, how fast was the car traveling before it attempted to stop? Round to the nearest whole number.

2. Based on the table, what percent of pedestrians die after being hit by a car moving at that speed?

3. Based on the cooling of the body, if the victim died instantly, how long ago did the accident occur? Round to the nearest hour.

4. Can you think of any other factors that should be taken into consideration before determining whether the impact of the car was the cause of death?

Math@Work
Other Careers in Mathematics

Earning a degree in mathematics or minoring in mathematics can open many career pathways. While a degree in mathematics or a field which uses a lot of mathematics may seem like a difficult path, it is something anyone can achieve with practice, patience, and persistence. Three growing fields of study which rely on mathematics are actuarial science, computer science, and operations research. While each of these fields involves mathematics, they require special training or additional education outside of a math degree. A brief description of each career is provided below along with a source to find more information about these careers.

Growing Fields of Study

Actuarial Science: The field of actuarial science uses methods of mathematics and statistics to evaluate risk in industries such as finance and insurance. Visit www.beanactuary.org for more information

Computer Science: From creating web pages and computer programs to designing artificial intelligence, computer science uses a variety of mathematics. Visit computingcareers.acm.org for more information.

Operations Research: The discipline of operations research uses techniques from mathematical modeling, statistical analysis, and mathematical optimization to make better decisions, such as maximizing revenue or minimizing costs for a business. Visit www.informs.org for more information.

There are numerous careers that have not been discussed in this workbook. Exploring career options before choosing a major is a very important step in your academic career. Learning about the career you are interested in before completing your degree can help you choose courses that will align with your career goals. You should also explore the availability of jobs in your chosen career and whether you will have to relocate to another area to be hired. The following web sites will help you find information related to different careers that use mathematics. Another great resource is the mathematics department at your college.

The **Mathematical Association of America** has a website with information about several careers in mathematics. Visit www.maa.org/careers to learn more.

The **Society for Industrial and Applied Mathematics** also has a webpage dedicated to careers in mathematics. Visit www.siam.org/careers to learn more.

The **Occupational Outlook Handbook** is a good source for information on educational requirements, salary ranges, and employability of many careers, not just those that involve mathematics. Visit http://www.bls.gov/ooh/ to learn more.

Answer Key

R.1 Exercises

1. False; equals 81
3. False; a prime number has exactly 2 factors.
5. Base = 2, Exponent = 9
7. 4
9. Composite Number, $1 \cdot 34$, $2 \cdot 17$
11. $3 \cdot 5 \cdot 7$
13. 120
15. 120 days

R.2 Exercises

1. False; the numerator is 11.
3. True
5. $\dfrac{1}{7} = \dfrac{5}{35}$
7. $\dfrac{5}{23}$
9. $\dfrac{7}{24}$
11. $\dfrac{27}{7}$
13. 100
15. No. If a fraction is less than 1 then its product with another number will be less than that other number. So, if the other number is less than 1, the product will be less than 1. Answers will vary.

R.3 Exercises

1. True
3. False; when subtracting fractions, subtract the numerators and keep the common denominator.
5. $\dfrac{5}{8}$
7. $\dfrac{2}{3}$
9. $\dfrac{7}{18}$
11. $3856
13. 1. Find the LCD. 2. Change each fraction to an equivalent fraction that has the common denominator. 3. Add or subtract the numerators and keep the common denominator. 4. Reduce if possible.

R.4 Exercises

1. False; on a number line, any number to the right of another number is larger than that other number.
3. False; the decimal points do not need to be aligned vertically when multiplying decimal numbers.
5. 82.072
7. one hundred eighty-seven and nine hundred forty-two thousandths
9. 89.82
11. 2.2
13. 0.69674
15. 28

R.5 Exercises

1. True
3. True
5. a. $42,752.00
   b. $54,817.00
   c. New Hampshire
   d. $12,065.00
   e. 93.56%
7. a. 25.3
   b. 27.5

1.1 Exercises

1. True
3. True
5. [number line with points at -3, -1, 1, 5]
7. Rational Number, Integer, Whole Number, Natural Number
9. True
11. $4.2 > -5.9$
13. No Solution $(\emptyset)$
15. -4500 meters
17. If y is a negative number then $-y$ represents a positive number. For example, if $y = -2$, then $-y = -(-2) = 2$.

1.2 Exercises

1. False; The sum of a positive and negative number can be positive, negative, or zero.
3. True
5. False; The mean of a set of numbers can be positive, negative, or zero.
7. -97
9. 60
11. 15
13. -6400 feet
15. 108 sq ft

1.3 Exercises

1. True
3. False; Quotient indicates division.
5. 6
7. 975 students
9. $\$929$
11. 8.2 degrees

1.4 Exercises

1. False; In the expression $\sqrt{81}$, the number 81 is the radicand.
3. True
5. 1
7. 9
9. -6
11. a. $0.10 \cdot 4 \cdot (\$102 + 7.50)$
    b. $\$394.20$
13. The radical sign is $\sqrt{}$. The radicand is the number under the radical sign. A radical expression includes the radical sign and its radicand. For example, in the expression $\sqrt{36}$ the entire expression is called a radical expression, 36 is the radicand, and the symbol $\sqrt{}$ is called the radical sign.

1.5 Exercises

1. False; The commutative property of addition allows the order to change.
3. False; The additive identity of all numbers is 0.
5. Commutative Property of Addition
7. Associative Property of Multiplication
9. Additive Identity Property
11. a. $\$3.99 + \$2.25 + \$12.01 + \$1.75$
    b. $(\$3.99 + \$12.01) + (\$2.25 + \$1.75) = \$16 + \$4 = \$20$
    c. Commutative and associative properties of addition
    d. Yes

1.6 Exercises

1. True
3. False; In the term "12a," 12 is the coefficient.
5. $-1.3x$
7. $5x + 0.4$
9. 36
11. $\$50,000$

13. A number is a constant and its value does not change regardless of where it appears in an expression or equation. A variable, usually represented by a letter, represents an unknown number or numbers.

1.7 Exercises

1. True
3. False; Subtraction is indicated by the phrase "five less than a number."
5. $x + 3$
7. $x(2y)$
9. The sum of three and a number
11. The commutative property of addition and multiplication permits the order of items being added or multiplied to change and still have the same result. This property does not hold true for subtraction or division. Therefore, order is important for subtraction and division problems or the answer will change or be incorrect.

2.1 Exercises

1. True
3. True
5. 2 is a solution
7. $w = -7$
9. -3
11. $x = 23$ boxes of dry erase markers
13. a. Yes. It is stating that $6 + 3$ is equal to 9.
  b. No. If we substitute 4 for x, we get the statement $9 = 10$, which is not true.

2.2 Exercises

1. False; Subtract 3 from both sides.
3. False; It is called a contradiction.
5. $z = -1$
7. $z = -2$
9. 4
11. Identity
13. 6250 tickets/hour
15. a. $5x + 1$
  b. $x = 6$
  c. Answers will vary.

2.3 Exercises

1. False; case matters in formulas.
3. True
5. $x = \dfrac{-9 + 8y}{3}$
7. 153 yards

2.4 Exercises

1. True
3. False; the first step is to read the problem carefully.
5. 53
7. $8500
9. 3.75 hours
11. $135
13. $10

2.5 Exercises

1. True
3. False; a proportion is a statement that two ratios are equal.
5. $\dfrac{5}{2}$
7. $\dfrac{14}{3}$

414 Answer Key   3.1 Exercises

11. False

13. $x = 117$

15. $897

2.6 Exercises

1. False; they vary directly.
3. True
5. $y = 256.15$
7. $z = 150.19$
9. 175 lb

11. a. When two variables vary directly, an increase in the value of one variable indicates an increase in the other, and the ratio of the two quantities is constant.
b. Joint variation is when a variable varies directly with more than one other variable.
c. When two variables vary inversely, an increase in the value of one variable indicates a decrease in the other, and the product of the two quantities is constant.
d. Combined variation is when a variable varies directly or inversely with more than one variable.
Answers will vary.

2.7 Exercises

1. True
3. False; only one value in the solution set needs to be checked.
5.

7. a. $(-\infty, -2]$
b.

9. a. $(-\infty, -2]$
b.

11. a. $(3, 7)$
b.

13. $83 < x \leq 100$
15. a. Answers will vary.
b. Answers will vary.

3.1 Exercises

1. True
3. False; Horizontal lines have *y*-intercepts.
5. a.

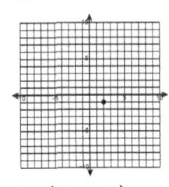

A: $(2, -1)$
b. A: $(-1, -4)$

7. a. $\dfrac{8}{5}$
b. $\dfrac{3}{5}$

9.

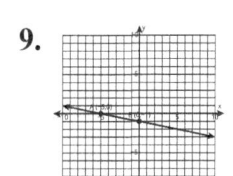

A: $(-5, 0)$; B: $(0, -1)$

11.

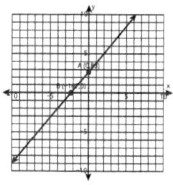

y-intercept (A): $\left(0, \dfrac{8}{3}\right)$; x-intercept (B): $\left(\dfrac{-16}{7}, 0\right)$

13. **a.** $(2, 8), (6, 24), (7, 28), (9, 36)$

b.

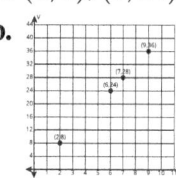

3.2 Exercises

1. True
3. True
5. $\dfrac{-3}{13}$
7. **a.** Horizontal, 0
b.

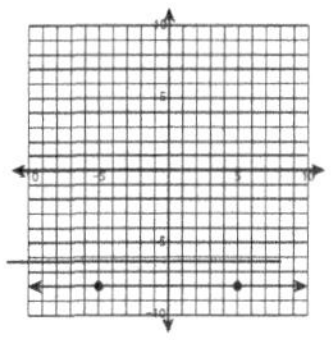

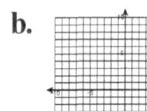

9. **a.** Slope: $\dfrac{-1}{2}$, y-intercept: $(0, 4)$
b.

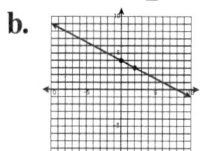

11. **a.** $y = 2x + 1$
b.

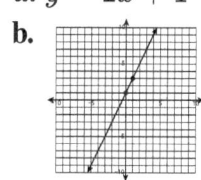

13. $-\$1250$ per year

3.3 Exercises

1. True
3. True
5. $y = -\dfrac{10}{9}x - \dfrac{7}{2}$
7. $y = 3x + 5$
9. $y + 2 = -\dfrac{2}{3}(x + 3)$

11.

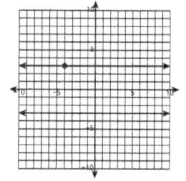

$y = 3$

13. **a.** $f = 5 + 2m$
b. $\$35$

3.4 Exercises

1. True
3. True

5. **a.** Domain: $0, -4, 7, 6, -7$, Range: $-8, 1, 2, 6, -2$
b. No

416 Answer Key 3.7 Exercises

7. **a.** Yes
   b. Domain: $[1, 7]$ or $1 \le x \le 7$, Range: $[-10, -7]$ or $-10 \le y \le -7$

9. $D = (-\infty, 2) \cup (2, \infty)$
11. $f(4.3) = 26.2$
13. $f(-2) = 2$

3.5 Exercises

1. True.
3. False; if $r = 1$ or -1 the variables are perfectly correlated.
5. a.
7. $y = -5x + 90$
9. $\hat{y} = -1 + 2x$

11. Answers will vary. For example, if the data is positively correlated, the line between one pair of points in the data set might go through the data set with a positive slope while the line between another pair of data points in the data set might have a negative slope and intersect with the data set.

3.6 Exercises

1. False; graphs that are parallel indicate no solutions.
3. True
5. Equation $x + y = 11$ in slope-intercept form:
   $y = -x + 11$
   Equation $-5x + 5y = 5$ in slope-intercept form:
   $y = x + 1$
   The system of equations is consistent and the equations are independent.
   The solution to this system of equations is $(5, 6)$.

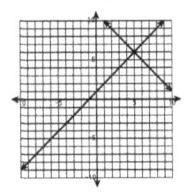

7. One Solution, $(0, 2)$
9. 10.59 and 11.41
11. **a.** Answers will vary.
    b. Answers will vary.

3.7 Exercises

1. True
3. False; The solution is the intersection of the graphs.
5. 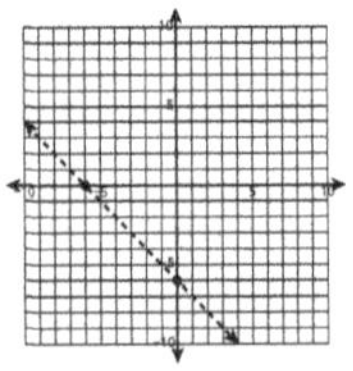
7. 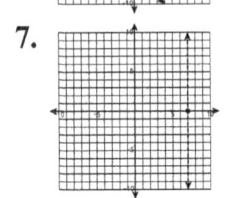
9. **a.** 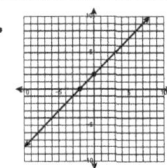
   b. 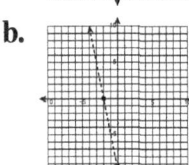
   c.

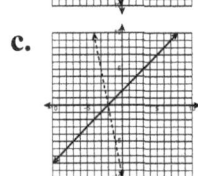

11. a. $\begin{cases} 0.1x + 0.3y + 1.2 < 5 \\ x + 2y < 30 \end{cases}$

b.

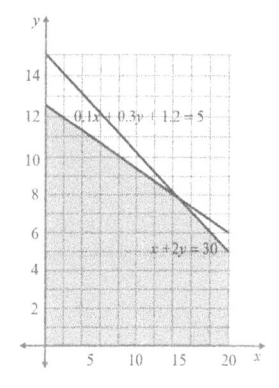

c. Answers will vary.

d. Answers will vary. For example, any negative values of *x* and *y* since you cannot have a negative number of cookies.

4.1 Exercises

1. False; if there is no exponent written, the exponent is assumed to be 1.
3. True
5. x^7
7. 1
9. x^8
11. $\dfrac{-4}{x^9}$
13. $\dfrac{x^8}{4}$
15. $\dfrac{1}{81x^{12}y^{16}}$
17. Positive

4.2 Exercises

1. True
3. False; 3.53×10^5 is greater than 8.72×10^{-4}.
5. 2.5×10^4
7. 2×10^{-2}
9. 3.25×10^{-19} grams

4.3 Exercises

1. False; they increase slowly at first and then grow very quickly.
3. True
5. Exponential model
7. $1197.04
9. $10,171.77

4.4 Exercises

1. False; it is a monomial with a degree of 0.
3. False; all terms are subtracted.
5. No
7. $-2x^2 + 11x - 12$
9. $4x + 3$
11. a. $f(3) = 37$
    b. $f(a) = 3a^2 + 3a + 1$
13. a. 213 protozoa after 5 hours
    b. 164,229 after 2 days (48 hours)

4.5 Exercises

1. False; the distributive property can be used when multiplying any types of polynomials.
3. False; the product will be a binomial.
5. $-6x^2 + 9x + 3$
7. $x^3 + 3x^2 - 6x - 8$
9. $6x^2 - 6$
11. $81x^2 - 1$
13. $81x^2 + 18x + 1$
15. $A(x) = 9x + 18$

5.1 Exercises

1. False; variables need to be considered as well.
3. True
5. 1
7. $-2 + \dfrac{2}{x}$
9. $7y(5x^2 - y)$
11. $3x^2y(5 + 11y + y^2)$
13. Not Factorable By Grouping
15. a. 32 feet
    b. $16x(3 - x)$
    c. 32 feet
    d. Yes. They are equivalent expressions.

5.2 Exercises

1. True
3. False; the first step is to multiply a and c.
5. $(x - 6)(x - 4)$
7. $(5x - 7)(x - 2)$
9. Not Factorable
11. This is not an error, but the trinomial is not completely factored. The completely factored form of this trinomial is $2(x + 2)(x + 3)$.

5.3 Exercises

1. True
3. False; The sum of two squares is not factorable.
5. $(3y + 2x)(3y - 2x)$
7. $(5x + 1)(5x + 1)$ or $(5x + 1)^2$
9. $(3y + 4x)(9y^2 - 12yx + 16x^2)$
11. $(8x + 7)(x - 1)$

5.4 Exercises

1. True
3. True
5. $x = 10, \dfrac{-8}{9}$
7. $x = -5, 0, 7$
9. $x = \dfrac{3}{4}$
11. a. 640 ft; 384 ft
    b. 144 ft; 400 ft
    c. 7 seconds; $0 = -16(t + 7)(t - 7)$
13. This allows for use of the zero factor property which says that for the product to equal zero one of the factors must equal zero. Answers will vary.

5.5 Exercises

1. False; if the original number is negative, the principal square root will not be the same as the original number.
3. False; $\sqrt{a}$ and $\sqrt[3]{a}$ do not have the same index, so they are not like radicals.
5. Not a Real Number
7. $\dfrac{6}{7}$
9. $2\sqrt{5}$
11. $18\sqrt{2} + 4\sqrt{5}$
13. $18 - 8\sqrt{6}$
15. $\dfrac{2\sqrt{11}}{11}$
17. $r = 9\sqrt{3}$ cm

5.6 Exercises

1. True
3. False; there are two real solutions.
5. $\dfrac{5}{3}, -\dfrac{5}{3}$
7. Two different real solutions, $\dfrac{2-3\sqrt{2}}{2}, \dfrac{2+3\sqrt{2}}{2}$
9. $1+\sqrt{3}, 1-\sqrt{3}$
11. a. 3.35 seconds
    b. 4.15 seconds
    c. 0.80 seconds
13. $x^4 - 13x^2 + 36 = 0$; multiplied

5.7 Exercises

1. False; the sum of the squares of the lengths of the legs is equal to the hypotenuse squared.
3. 5 and 10 or -10 and -5
5. Width $= 9$ and Length $= 15$ or Width $= 15$ and Length $= 9$
7. 6 feet by 8 feet
9. a cannot be equal to zero and b^2 must be greater than or equal to $4ac$ to produce a real solution. Also, all the solutions found may not apply to the problem at hand. You must check that each answer makes sense in the context of the problem.

5.8 Exercises

1. False; the vertex is either the highest or lowest point on the parabola.
3. False; quadratic functions of the form $y = ax^2 + bx + c$ have a line of symmetry at $x = -\dfrac{b}{2a}$
5. a. $(2, 5)$
   b. $x = 2$
   c. Two: $\left(2-\sqrt{5}, 0\right), \left(2+\sqrt{5}, 0\right)$
   d. $(0, 1)$
   e. $(-1, -4), (3, 4)$
   f.
7. a. Time $= 3.16$ seconds
   b. Max Height $= 172.39$ ft
9. 51 ft by 102 ft

6.1 Exercises

1. True
3. True
5. 11.25 ft
7. 98 pt
9. $93.22
11. Colby would need to know that there are 3 feet in a yard and 5280 feet in a mile.

6.2 Exercises

1. False; to change from smaller units to larger units, division must be used.
3. False; in metric units, a square that is 1 centimeter long on each side is said to have an area of 1 square centimeter.
5. 466 900 cm
7. 9 090 000 cm^2
9. 1750 railroad ties

6.3 Exercises

1. True
3. False; in 1 liter there are 1000 milliliters.
5. True
7. milliliters
9. 0.049 kg
11. 60 doses
13. You could change to milliliters by multiplying by 1000 or by using a unit fraction where the numerator is 1000 mL and the denominator is the given measure in liters.

6.4 Exercises

1. False; water freezes at 32 degrees Fahrenheit.
3. False; a 5k (km) run is shorter than a 5 mile run.
5. 86 °F
7. 354.75 cm^2 or 354.839 cm^2
9. 239.2 yd^2 or 239.234 yd^2
11. 5.812 gal
13. 0.353 oz
15. 22 min

6.5 Exercises

1. True
3. True
5. acute
7. 8°
9. a. 51°
   b. 90°
   c. ∠2 and ∠5
11. straight angle

6.6 Exercises

1. True
3. False; $\tan \theta = \dfrac{\text{opp}}{\text{adj}}$.
5. Scalene
7. a. $m\angle Z = 80°$
   b. Acute
   c. $\overline{YZ}$
   d. $\overline{XZ}$ and $\overline{XY}$
   e. No, no angle is 90°
9. a. $\dfrac{75}{4}$
   b. $\dfrac{45}{4}$
11. 5 feet
13. Answers will vary. For example, you could use $\sin \theta$ to find the length of the side opposite θ and use $\cos \theta$ to find the length of the side adjacent to θ.

6.7 Exercises

1. a. True
   b. False; not all rectangles have four equal sides.
3. True
5. 96 m
7. 49 ft
9. 56.31 m
11. 95 ft^2
13. 1017.36 m^2
15. a. 320 ft^2
    b. 40 pounds

6.8 Exercises

1. True
3. True
5. 2688 ft^3
7. 2110.08 cm^3
9. 4056 cm^2
11. 800 ft^3
13. Volume is measured in cubic units. Volume takes up a three-dimensional space and the units can be thought of as small cubes which leads to the concept of cubic units.

6.9 Exercises

1. True
3. False; $\tan\theta = \dfrac{\text{opp}}{\text{adj}}$
5. $\dfrac{3}{5}$
7. 0.5
9. 346.4 feet

7.1 Exercises

1. False; to change a decimal number to a percent, move the decimal point two places to the right and add the % sign.
3. True
5. 1.2%
7. $\dfrac{1}{80}$
9. 209.79
11. 44.37%
13. 7%
15. a. He lost 10% of his original weight.
    b. He gains back 1119% of his weight
    c. The percentages are different because the gain of 20 pounds is based on his new weight of 180 pounds, not his original weight of 200 pounds.

7.2 Exercises

1. True
3. False; compound interest is earned on the principal and interest earned.
5. $78.75
7. $5000.00
9. $21,173.17
11. $461,000
13. Interest is money paid for the use of money, whether it is simple or compounded. Simple interest is derived using a formula that calculates the interest based on the starting principal one time over any amount of time. Compound interest not only earns interest on the principal but also on the interest itself. It can be a series of simple interest calculations where the principal changes each time to include the interest earned.

7.3 Exercises

1. False; the finance charge is the amount of interest paid over the life of the car loan.
3. True
5. $553.81
7. $5383.76
9. 8.5%
11. Answers will vary. For example, an oil change can cost between $20 to $70, and replacing a thermostat can cost between $100 and $300.

7.4 Exercises

1. True
3. False; a general guideline is to spend at most 30% of your income (before taxes) on housing costs.
5. down payment: $24,000; amount financed: $136,000
   down payment: $24,000; amount financed: $136,000
7. a. $104,000
   b. $29,980

9.
Payment Number	Interest Payment	Principal Payment	Mortgage Balance
1	$666.67	$288.16	$199,711.84
2	$665.71	$289.12	$199,422.72
3	$664.74	$290.09	$199,132.63

8.1 Exercises

1. False; the three ways to write sets are word descriptions, set-builder notation, and roster notation.
3. False; given $A = \{1, 2, 3, 4, 5\}$, A has $2^5 = 32$ subsets.
5. $A = \{11, 13, 15, 17, 19\}$
7. $A = \emptyset$ or $A = \{\ \}$
9. 7 proper subsets;
$\emptyset, \{6\}, \{8\}, \{14\}, \{6, 8\}, \{6, 14\}, \{8, 14\}$

8.2 Exercises

1. True
3. False; they have no elements in common.
5. $B = \{17, 29, 35, 54\}$
7. $A \cup B = \{1, 2, 3, 4, 5, 6, 7, 8, 10, 12\}$
9. a.
11. $A' = \{$b, c, d, e, f, g, h, i, j, k, l, m, n, p, q, r, s, v, w, x, y, z$\}$
13. Dan, Max

8.3 Exercises

1. True
3. False; it is an example of inductive reasoning.
5. 15
7. Deductive reasoning
9. b.

8.4 Exercises

1. True
3. True
5. This milk has no chocolate in it.
7. Some of the dogs are barking.
9. ~p: Five is not greater than three.
11. Answers will vary.

8.5 Exercises

1. False; compound statements contain logical connectives.
3. True
5. a. compound
   b. simple
   c. compound
7. Disjunction
9. $t \wedge r$
11. Flowers bloomed in May and it rained in April.
13. Answers will vary
   a: It is raining.
   b: The dog is muddy.
   c. The dog will get a bath.
   d. The dog will go back outside.
   $a \Rightarrow b$. $c \Leftrightarrow b$. $c \wedge \sim d$.

8.6 Exercises

1. True
3. True

5.

Truth Table		
w	z	$w \wedge z$
T	T	T
T	F	T
F	T	T
F	F	F

7.

Truth Table				
a	b	c	$a \wedge b$	$(a \wedge b) \Rightarrow c$
T	T	T	T	T
T	T	F	T	F
T	F	T	F	T
T	F	F	F	T
F	T	T	F	T
F	T	F	F	T
F	F	T	F	T
F	F	F	F	T

9. a: Tom is on the beach;
b: Tom is wearing sunscreen;
c: Tom will get a sunburn;
Statement: $(a \wedge \sim b) \Rightarrow c$

9.1 Exercises

1. False; the possible outcomes are a head or a tail.
3. True
5. { HHH, HHT, HTH, HTT, THH, THT, TTH, TTT }
7. $\frac{289}{1878}$ or 0.1539
9. $\frac{1}{6} \approx 0.166667$
11. Your friend, because the law of large numbers says that the greater the number of trials, the closer the empirical probability will be to the classical, or actual probability.

9.2 Exercises

1. False; they are not mutually exclusive.
3. True
5. Mutually Exclusive
7. $\frac{1}{12} \approx 0.083333$
9. 0.25
11. $\frac{4}{5} = 0.8$ points

9.3 Exercises

1. False; the probability of an event B, given that event A has occurred is usually not equal to the probability of an event A, given that event B has occurred.
3. False; the probability that mutually independent events occur at the same time is the product of their probabilities.
5. $\frac{4}{51}$ or 0.0784
7. $\frac{13}{102}$ or 0.1275
9. Answers will vary. For example, if A and B are dependent, then the occurrence of A will affect the probability of B, which means that we would have to use $P(B|A)$ instead of $P(B)$ in the calculation of $P(A \text{ and } B)$.

9.4 Exercises

1. False; there are ways for the two events to occur in the given order.
3. False; each ordering of a set of elements is called a permutation.
5. 6720
7. 60
9. 24
11. 120

13. Answers will vary. For example, there are 6 possible outcomes for the first digit. Whatever digit is used cannot be used again, so there are 5 possible outcomes for the second digit. Similarly, there are 4 possible outcomes for the third digit and 3 possible outcomes for the last digit. By the fundamental counting principle, we need to multiply each number of possible outcomes: $6 \cdot 5 \cdot 4 \cdot 3$.

9.5 Exercises

1. True
3. False; it is symbolized $_nC_r$.
5. **a.** This is a combination problem because the order in which the employees are picked doesn't matter.
b. Since the trophies are not all the same, order does matter. Therefore, this is a permutation.

7. 20
9. Answers will vary. For example,
$$_nC_r = \frac{_nP_r}{r!} = \frac{\frac{n!}{r!(n-r)!}}{\frac{r!}{1}} = \frac{n!}{(n-r)!} \cdot \frac{1}{r!} = \frac{n!}{r!(n-r)!}.$$

9.6 Exercises

1. $\frac{3}{8} = 0.375$
3. $\frac{1}{182} \approx 0.005495$

5. Answers will vary. Any scenario where order is important could use permutations while any scenario where order does not matter could use combinations.

10.1 Exercises

1. False; a census collects data from every member of a population.
3. True

5. 1000 of the 1883 participants in last year's Ironman in Kona, Hawaii were interviewed. 68% of those interviewed said they became interested in triathlons after first competing in marathons.
7. Cluster
9. Answers will vary.

10.2 Exercises

1. True
3. False; the next class should be 19–22.

5.

Grade	Frequency
A	1
B	2
C	7
D	4
F	7

7. **a.** 8
b. 6
c. 33
9. **a.** 112
b. 21
c. 187

11. Answers will vary. When creating a grouped frequency distribution, split the data so that all the classes are the same size and there is no overlap. If necessary, extend the range of the last class so that all class widths are the same.

10.3 Exercises

1. True
3. True
5. a. 9
   b. 11
   c. No Mode
7. 5.561
9. a. 6.9
   b. 7
   c. 6

10.4 Exercises

1. False; the range takes only two data values into consideration.
3. False; quartiles divide a data set into four equal parts.
5. 19
7. a. $Q_2 = 64.5$
   b. $Q_1 = 54$
   c. $Q_3 = 81$
9. 134.58
11. Answers will vary. For example, Data Set 1: 2, 4, 5, 6, 7, 9, 10, 12; Data Set 2: 2, 5, 5, 6, 7, 7, 7, 12 The range for both data sets is 10. The standard deviation for data set 1 is 3.31, and the standard deviation for data set 2 is 2.83.

10.5 Exercises

1. True
3. False; the normal distribution only has one mode, and it is the center.
5. 68%
7. 16%
9. 0.13
11. 2.89

A.1 Exercises

1. False; it is read "b sub four three".
3. False; it is written as the number of rows by the number of columns.
5. 3 x 3 matrix
7. $\begin{bmatrix} 6 & -4 & 8 \\ -7 & -5 & 15 \end{bmatrix}$
9. The matrices cannot be subtracted.
11. $\begin{bmatrix} -56 & 56 \\ -32 & -24 \\ -24 & -48 \end{bmatrix}$

Notes

Notes

Notes

Notes